Narrative SQL

Crafting Data Analysis Queries That Tell Stories

Hamed Tabrizchi

Apress®

Narrative SQL: Crafting Data Analysis Queries That Tell Stories

Hamed Tabrizchi ⓘ
University of Tabriz
Tabriz, Iran

ISBN-13 (pbk): 979-8-8688-1559-1   ISBN-13 (electronic): 979-8-8688-1560-7
https://doi.org/10.1007/979-8-8688-1560-7

Managing Director, Apress Media LLC: Welmoed Spahr
Acquisitions Editor: Shaul Elson
Development Editor: Laura Berendson
Coordinating Editor: Gryffin Winkler
Copy Editor: Kezia Endsley

Cover image by Christian Horz from stock.adobe.com

Distributed to the book trade worldwide by Springer Science+Business Media New York, 1 New York Plaza, New York, NY 10004. Phone 1-800-SPRINGER, fax (201) 348-4505, e-mail orders-ny@springer-sbm.com, or visit www.springeronline.com. Apress Media, LLC is a Delaware LLC and the sole member (owner) is Springer Science + Business Media Finance Inc (SSBM Finance Inc). SSBM Finance Inc is a **Delaware** corporation.

For information on translations, please e-mail booktranslations@springernature.com; for reprint, paperback, or audio rights, please e-mail bookpermissions@springernature.com.

Apress titles may be purchased in bulk for academic, corporate, or promotional use. eBook versions and licenses are also available for most titles. For more information, reference our Print and eBook Bulk Sales web page at http://www.apress.com/bulk-sales.

Any source code or other supplementary material referenced by the author in this book is available to readers on the Github repository. For more detailed information, please visit https://www.apress.com/gp/services/source-code.

If disposing of this product, please recycle the paper

To my adorable father, Hamid, my kind-hearted mother, Soheyla, and my wonderful brother, Mohammad—each of whom helped me understand the true value of life.

To Shaul and Gryffin, who believed in me and stood by me every step of the way.

Table of Contents

About the Author

Hamed Tabrizchi is an experienced data analyst and engaging storyteller who has more than five years of experience turning complex data into compelling narratives. Passionate about educating others, Hamed lectures at universities, leads workshops, and contributes to leading scientific journals. As a result of his observations in the professional community, he decided to write this book to fill a void he observed—the need for a resource that weaves SQL technicalities with narratives to empower analysts in delivering insights that resonate and drive action.

About the Technical Reviewer

 **Alexander Arvidsson** is the chief technology officer at Analytics Masterminds, where he spends his days helping clients of all shapes and sizes take better care of—and make more sense of—their data.

He has spent the last 25 years poking around with data, databases, and related infrastructure services such as storage, networking, and virtualization, occasionally emerging from the technical darkness to attend a *Star Wars* convention somewhere in the world.

He is a long-time data platform MVP, frequent international speaker, podcaster, Pluralsight author, blogger, and a Microsoft Certified Trainer, focusing on the Microsoft data platform stack.

Acknowledgments

Writing a book is often seen as a solitary endeavor, but this one would not exist without the support, encouragement, and generosity of so many people.

My deepest thanks go to Apress for placing their trust in me as a first-time author and for accepting my proposal to write this book. I would like to thank Shaul Elson for his time, guidance, and belief in me. I'm also sincerely grateful to Gryffin Winkler for his dedication and support throughout this process. Special thanks to Alexander Arvidsson, whose valuable feedback, sharp eye, and honest critique—shared chapter by chapter— have shaped this work more than words can express.

This journey would not have been possible without the patience, love, and unwavering belief of my family.

Finally, to the readers—thank you for making space in your analytical and curious minds for these words.

Introduction

In the past decade, data analysis and SQL have played a central role in my professional career. My passion for query writing began during my bachelor's studies, where I completed a database course with full marks. As a result of this achievement, I was given the opportunity to become a teaching assistant the following semester, where I gained experience writing queries and explaining them to students. Despite speaking with a trembling voice at first, this role helped me gain confidence in technical communication and public speaking, and enhanced my advanced query-writing skills. These early experiences laid the foundation for my current expertise in SQL and data analysis, fundamental to my next career accomplishment.

After a year, I started working for a technology company, where I encountered more complex challenges. As a data analysis intern, two of the challenges I encountered included the lack of neatly organized data and the difficulty of collaborating. It was difficult and very different from teaching or solving textbook exercises to coordinate projects among programmers, query writers, UI/UX designers, and the other members of the team. Despite all the challenges, I was motivated day by day to gain experience and skills from my colleagues and improve my analytical skills to become a data analyst who has an insightful perspective.

Throughout the years, I have dealt with a number of projects and gained deeper insights and better analytical skills from the past. One day I decided that I had an insight that I could share with those who are interested in data analysis and query writing. So, I decided to write this book to teach what formed this perspective within me, in as much detail as I could. The core concept of this book is SQL query writing, which is at the core of my day-to-day activities, whether as a data analyst, a university lecturer, or a data team leader.

I believe that at the present time, people who are able to shape information into compelling stories hold an advantage in the ever-increasing world of data. Due to this belief, narrative SQL emerged, which is the idea that learning SQL should not be like learning a machine's language, but should instead feel like mastering a language of communication.

This book is for the curious analyst, the thoughtful developer, and the future storyteller of data. This book is designed to provide you with clear and creative guidance regardless of whether you are just beginning your journey into databases or want to improve your proficiency in SQL.

Using a narrative structure, this book begins with the basics—simple SELECT statements, filters, and JOINs. In the subsequent chapters, this book explores queries that transform raw data into rich insights, including aggregations, subqueries, conditional logic, and more. With the SQL queries and stories provided, this book is not just a reference guide; it's a companion for your data journey, helping you think narratively, write clearly, and analyze clearly.

Each chapter explores stories that introduce concepts and skills toward mastery of powerful SQL tools, including window functions, subqueries for dynamic data manipulation, conditional logic with complex queries, and even optimization strategies based on indexes and views. Upon completion of this book, you should be able to tackle a wide range of data analysis challenges by writing SQL queries. The last few chapters of this book cover advanced topics such as tuning performance, optimizing scripts, and analytical storytelling with window functions, giving your narratives depth and precision. In this book, you will find both inspiration and practical skills—and when you close the last chapter, you will be prepared to tell your own powerful data stories.

Finally, it should be noted that all queries presented in this book have been developed and thoroughly tested on PostgreSQL 14.17, the enterprise-grade open-source relational database system known for its robustness, extensibility, and SQL compliance. Although the fundamental concepts of PostgreSQL should apply to all PostgreSQL versions, specific syntax, performance characteristics, or feature availability might be different.

The complete collection of queries, including stories and examples, can be accessed via the publisher's GitHub repository at https://github.com/Apress/Narrative-SQL. Throughout this repository, all queries are conveniently organized and categorized by chapter, allowing you to find and execute examples relevant to specific sections conveniently.

Chapter Overviews

Chapter 1: The Storyteller's Database

The purpose of this chapter is to provide a foundation for your journey into the world of data and narrative. This chapter introduces databases as storytelling tools, illustrating how narrative structures and relational models can aid in making data meaningful. As you go through this chapter, you are provided with all the information you need to get started on this journey by setting the foundation for the art and science of data storytelling. You will learn that SQL is not only capable of querying data, but it can also tell compelling stories. In the next step, you will begin to explore SQL's complexities in greater detail after setting up the storytelling environment.

Chapter 2: Starting with SELECT

In this chapter, you learn how to extract and explore basic data from tables using the SELECT statement. Using SQL's most commonly used command, you can retrieve and manipulate data effectively. This requires you to learn how to write precise SELECT statements in order to retrieve information that is needed for your narratives. This will set the stage for more advanced data manipulation and analysis techniques.

Chapter 3: Filtering Facts with WHERE

In this chapter, you discover another SQL command that is frequently used to refine data retrieval by using WHERE conditions, comparisons, and logical operators. This requires creating precise WHERE statements that filter data based on specific criteria. This will enable you to refine your datasets and extract even deeper insights, which in turn helps you tell richer stories with your data.

Chapter 4: Complex Characters with JOINs

The purpose of this chapter is to explore how to connect multiple tables using JOINs, creating richer data narratives based on different sources of data. JOIN operations, which are fundamental to combining data from multiple tables, are discussed. As you become proficient in this operation, you will be able to create complex queries that provide deeper insights and more comprehensive analyses of your records.

Chapter 5: Aggregating Acts

The purpose of this chapter is to introduce aggregate functions such as COUNT, SUM, AVG, MIN, and MAX, which can be used to summarize and analyze grouped data. In SQL, an aggregate act is the application of aggregate functions to grouped data subsets. These actions enable SQL to extract useful summary and statistical information from data for analysis and decision-making.

Chapter 6: Ordering the Plot with ORDER BY and LIMIT

In this chapter, you learn how to sort query results and limit output in order to improve readability and performance. The focus is on sorting and filtering query results efficiently. Once you have mastered this operation, it will be possible to organize data meaningfully. To focus on the most relevant data points, you can prioritize key information and limit the results to the most relevant data points.

Chapter 7: Dynamic Dialogues with Subqueries

The purpose of this chapter is to present subqueries as powerful tools for nesting logic and constructing complex, layered data requests. This chapter explores the art of writing subqueries in order to add depth and dimension to data analysis. This chapter provides an overview of subqueries, their types, and narrative examples of their use in dynamic dialogues.

Chapter 8: Conditional Logic in Data Plotting

This chapter explains how SQL's conditional logic can be used to transform data analysis and visualization workflows to enable logic-based data visualization and transformation. You learn about conditional logic in SQL, categorize data, apply dynamic filtering to improve plot relevance for enhanced visualizations, create color-coded data for visualizations, aggregate data using conditional expressions, and handle missing data in visualizations. The focus of this chapter is not on how to visualize or plot data, but on the crucial process of preparing data. In this chapter, SQL is used to manipulate, clean, and structure data before it is visualized.

Chapter 9: Optimizing Your Script with Indexes and Views

This chapter sheds light on how indexes can be used to improve query performance and how views can be used to simplify logic. Optimizing SQL queries can significantly improve performance when dealing with large data volumes. Indexes and views are both powerful tools for achieving this type of optimization. An index enables the database engine to locate rows more efficiently, thereby reducing the need to scan entire tables in order to retrieve data. Alternatively, views simplify complex queries by storing reusable SQL logic, improving readability and maintenance.

Chapter 10: Analytics Alchemy: Turning Data into Gold

Turning data into gold with SQL requires mastering advanced analytical functions that help you extract deeper insights from raw data and transform them into compelling narratives. The SQL language provides powerful functions that can be used to transform raw data into compelling narratives. The chapter also discusses how raw data can be transformed into a compelling story and how recursive queries can be used to structure query logic effectively.

Chapter 11: The Grand Finale: Presenting Your Data Story

In this chapter, your journey is nearing its end. Through chapter-by-chapter learning, you learned how to extract deep insights and information from raw data to address complex and advanced analytical questions using raw data. This chapter summarizes the previous chapters and provides insight into presenting a narrative for data analysis.

Appendix A: SQL Syntax Reference Guide

This appendix provides a quick-access syntax guide for common SQL statements and clauses.

Appendix B: Glossary of Terms

This appendix defines key terms used throughout the book for quick reference and deeper understanding.

Appendix C: PostgreSQL Elements Reference

This appendix highlights PostgreSQL-specific features by providing a comprehensive alphabetical list of SQL statements, clauses, operations, and functions available in PostgreSQL.

CHAPTER 1

The Storyteller's Database

This chapter provides the basis of your journey into the world of data and narratives. Beginning with the basics of data, databases, and data analysis, this chapter explores Database Management Systems (DBMS) and SQL's pivotal role in navigating these repositories. Through an exploration of SQL commands and the use of data types, you can tell powerful stories. This chapter provides you with the essentials you need to get started on this journey. It covers the art and science of data storytelling.

Introduction to Data

In today's society, data is the foundation upon which everything is built, and it impacts every aspect of our lives. There are countless sources of data available to us, from weather patterns tracked by satellites to the number of steps you take each day. The term *data* refers to the qualitative or quantitative attributes of a variable. A great deal of data is collected, observed, or created for the purpose of analyzing it and making decisions based on it. It is possible to store structured or unstructured data, ranging from numbers, text, and multimedia to complex datasets used in computing and research.

In a nutshell, data is the raw material of information, the basis for understanding the world and making informed decisions. Data on its own is like a pile of unrefined oil. In spite of the fact that data has value, it's useless until it's processed and analyzed. There is a great deal of value in data because it is capable of revealing hidden patterns, trends, and insights. The ability to analyze data allows people to make better decisions, solve complex problems, and drive innovation in business. It is widely accepted that data has become a part of every aspect of our lives in the digital age, and that it is the basis of all decision-making across sectors such as the healthcare, finance, and technology industries. Table 1-1 provides five fundamental questions and answers when exploring data.

© Hamed Tabrizchi 2025
H. Tabrizchi, *Narrative SQL*, https://doi.org/10.1007/979-8-8688-1560-7_1

Table 1-1. *Five Fundamental Questions and Answers for Exploring Data*

#	Question	Answer
1	Why is data so valuable?	The use of data leads to decisions, innovation, and progress. Data contains the fundamental insights and evidence required to make informed decisions, resolve complex problems, and predict future trends in an era in which information is power.
2	Who relies on data?	All of us. Data is used in a wide range of things, from businesses using customer data to modify their products, to governments using data to plan policies, to scientists using research data to make discoveries. Many sectors and societies rely on data.
3	Where does data come from?	The world over. This includes countless devices and sensors in the vast expanse of the Internet and the billions of devices and sensors making up the Internet of Things (IoT). From the depths of the oceans using climate monitoring equipment, to the far reaches of space with satellites collecting data about our universe. Each source of data provides unique insights that contribute to our collective understanding.
4	When is data used?	Every day, continuously. Data is used constantly throughout our lives and is not limited to specific moments. Data is used to guide emergency responses and to facilitate financial transactions, and it is used to influence long-term policy decisions and scientific research. Data has relevance that extends across timescales, from the immediate to the generational.
5	What does data do?	In data, transformation occurs. Data has the power to change the world in many ways. These include informing policy, driving economic growth, advancing science and healthcare, improving education, and enriching cultural understanding. By analyzing data, we can uncover patterns, predict outcomes, personalize experiences, and foster innovation.

Data Analysis

Similar to oil in the 20th century—which powered economies, revolutionized transportation, and played a fundamental role in industrial advancement—data is the fundamental resource driving societal progress, economic growth, and innovation in the 21st century. In the same way that oil must be extracted, refined, and distributed to use its energy, data must be collected, organized, and analyzed to unlock its full potential. Thus, the importance of controlling the extraction, refinement, and analysis of data in this era cannot be understated. The purpose of data analysis is to find useful information, provide insight into conclusions, and support decision-making through inspection, cleansing, transforming, and modeling the data.

Databases

Databases are structured collections of data that are stored electronically and accessed by a computer system. This system facilitates the efficient organization, management, and retrieval of data. A database is designed to handle large amounts of data, allowing users to add, modify, and query data quickly and securely. Databases support a variety of data types, including text, numbers, multimedia files, and more, organized in a manner that facilitates the analysis of business operations, the management of transactions, and decision making. Thousands of applications rely on databases, from the websites people visit daily to the financial systems that operate on a global scale.

There is more to databases than just storing data; they are intricately designed to organize information in a way that makes it easily accessible and useful. Organization is of considerable importance, as the value of data lies not only in its existence, but in the ability to retrieve and interpret it. In every sector of society—be it healthcare, education, business, or technology—databases help manage patient records, student information, financial transactions, and more.

In general, there are several types of databases, including structured, semi-structured, and unstructured. As a result of the need to optimize storage, retrieval, and analysis of data based on its nature, there is a relationship between data types and database types. Structured, semi-structured, and unstructured data all require different database features. For analysis, it is crucial to be aware of the distinctions between each, since each requires different tools and approaches in order to extract its insights.

In structured data, each piece of information is organized and formatted in a way that is easily searchable in databases and spreadsheets, and it is stored in predefined models or schemas. In this way, computers can perform efficient processing and analysis. Structured data can be easily accessed and analyzed through relational databases.

In the real world, data is not always stored in a structured form and may be unstructured or semi-structured. Unstructured data consists of everything from emails to videos to social media posts, often stored in non-relational (NoSQL) databases that handle diverse and dynamic datasets. Semi-structured data straddles the line, combining elements of both. Examples of semi-structured data are JSON and XML documents, which, although not fitting into traditional table schemas, have inherent structure that can be queried and analyzed.

Relational Databases vs Non-Relational Databases

There is a major distinction between relational and non-relational databases, each with its own characteristics, advantages, and applications. A relational database is based on the relational model of data. A table (relation), which consists of rows and columns, is used to organize data in this model. A row represents a unique record, and a column represents a field. The power of relational databases lies in their use of SQL (Structured Query Language) for data manipulation and retrieval, which allows high levels of flexibility and precision in querying the data. The most popular relational database management systems (RDBMS) are PostgreSQL, MySQL, Oracle Database, and Microsoft SQL Server. Alternatively, non-relational databases, known as *NoSQL* databases, emerged as a response to the limitations of relational models, especially in handling large volumes of unstructured or semi-structured data. As opposed to having fixed schemas, these databases are often capable of storing a variety of data types, such as documents, key-values, wide columns, and graphs. NoSQL databases are particularly well suited to applications that require rapid development, scalability, and the ability to handle a variety of data types. Among the most popular are MongoDB (document-based), Redis (key-value store), Cassandra (wide-column database), and Neo4j (graph database).

Transferring between structured, semi-structured, and unstructured data types is often driven by a variety of needs and challenges in data management and analysis. Each has its unique characteristics and optimal use cases. Thus, a variety of databases and database types exist, ranging from relational databases, which can handle structured data effectively through well-defined schemas and relationships, to NoSQL databases such as document, key-value, wide-column, and graph databases. They are each tailored to meet the specific requirements of unstructured or semi-structured data. As the title of the book indicates, the following chapters focus on structured data, exploring the expansive world of relational databases and the use of the SQL language to interact with them.

Exploring Relational Database Management Systems (RDBMS)

Relational databases are crucial throughout the entire process. As relational databases are structured, they ensure the integrity and consistency of data, which is crucial for data analysis. The powerful querying capability of SQL allows analysts to retrieve specific subsets of data quickly and efficiently from large databases. SQL's querying ability enables analysts to perform complex aggregations, joins, and filtering operations with ease, which are essential tasks in the data preprocessing and exploration phases. The purpose of this book is to teach you how to extract and analyze data stored in databases using SQL. Throughout the remainder of this chapter, you learn more and more about data analysis, databases, and other concepts, but you first need a better understanding of the data in order to be able to analyze it.

In most relational databases, SQL is used to query and manage the data. Due to SQL's powerful and flexible capabilities, it has become the standard language for relational database management systems (RDBMS). There are a number of popular relational database systems that use SQL, including MySQL, PostgreSQL, Oracle Database, Microsoft SQL Server, and SQLite. Due to the standard way in which SQL interacts with structured data, these systems can perform complex queries, update data, create and modify schemas, and manage database access more easily.

This book uses PostgreSQL as the basis for its SQL examples, which offers numerous advantages for readers who are eager to gain a deeper understanding of data analysis through the lens of relational databases. The PostgreSQL database system is well known for its robustness, open-source nature, and compliance with SQL standards, making it one of the most advanced and reliable relational database systems. Through its open-source model, users are not only able to access a high-quality database system without licensing fees, but also benefit from a vibrant developer community that is constantly enhancing its capabilities. In addition to complex SQL queries, foreign keys, triggers, views, and stored procedures, PostgreSQL supports a wide range of SQL functionality.

Databases in Data Analysis and Storytelling

Databases provide more than just a means of storing data; they are also instrumental in analyzing data and telling stories based on that data. In data analysis, data is examined, cleaned, transformed, and modeled in order to find useful information, draw conclusions, and support decision-making. By structuring and organizing data, databases facilitate this process; they enable analysts to query and manipulate data effectively. On the other hand, storytelling involves using narratives to communicate information in an engaging and understandable manner. Data storytelling involves crafting narratives around insights in data in order to make complex information accessible and understandable. The definition of databases in the context of data storytelling could be expressed as a source of truth from which narratives are constructed in data storytelling. As a result of enabling the extraction of meaningful patterns and trends, databases make it easy for storytellers to tell narratives that are in tune with their audiences. In a nutshell, databases, by providing the raw material for these stories, lie at the heart of this process.

Diving into SQL

In the 1970s, the creation of SQL marked a pivotal moment in the evolution of data storage and retrieval. Over the decades, SQL evolved from a simple query language to a tool for professionals working with data. The journey of SQL began in the early 1970s at IBM, where researchers Donald D. Chamberlin and Raymond F. Boyce developed a prototype called SEQUEL (Structured English Query Language). This prototype was designed to manipulate and retrieve data stored in IBM's early relational database management system. The language was late renamed SQL to avoid brand-name issues. By the late 1970s and early 1980s, SQL had been adopted as the standard language for RDBMSs. Since then, SQL has undergone several revisions to include updated features and capabilities. These features include support for XML data, window functions, and expanding its utility and efficiency in managing diverse data types and complex queries. SQL allows users to interact with databases to perform operations such as querying, updating, inserting, and deleting data.

SQL Command Types: The Five Principles of Database Interaction

There are five distinct types of commands in SQL, each of which performs a specific function when it comes to managing and manipulating data. These categories are Data Definition Language (DDL), Data Manipulation Language (DML), Data Control Language (DCL), Transaction Control Language (TCL), and Data Query Language (DQL).

- **Data Definition Language (DDL)**: DDL commands are used to define, alter, and manage the schema and structure of database objects like tables, indexes, and views. These commands do not manipulate the data itself but instead shape the "containers" that hold the data, allowing for the creation and modification of database structures.

- **Data Manipulation Language (DML)**: DML commands are likely to be the most frequently used, as they deal directly with data manipulation within existing database structures. They enable users to insert, update, delete, and manage the database data.

- **Data Control Language (DCL)**: DCL commands are focused on permissions and access control for database objects. For security and confidentiality, these commands are crucial in multi-user databases.

- **Transaction Control Language (TCL)**: TCL commands manage the changes made by DML operations as transactions, which are either completely processed or not processed at all, ensuring data consistency and integrity. These commands allow users to commit or roll back changes to the database.

- **Data Query Language (DQL)**: DQL deals with retrieving data and is primarily represented by the SELECT command, which queries data from tables within a database. DQL allows users to specify exactly which data should be returned from the query, making it a powerful tool for extracting and analyzing information stored in the database.

Figure 1-1 illustrates an overview of the SQL command types, indicating each category along with its respective commands. This illustration serves as a guide, mapping out the distinct SQL command types. This visualization not only aids in understanding the functional divisions within SQL but also highlights the specific operations that can be performed within each category.

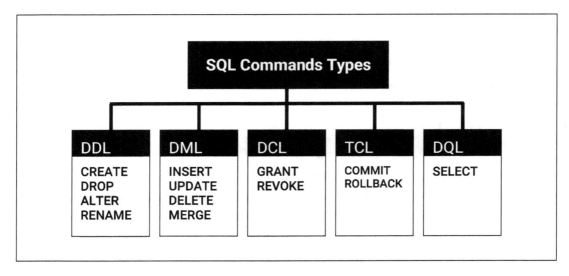

Figure 1-1. *SQL command types*

When it comes to data analysis, you will mainly be working with Data Manipulation Language (DML) and Data Query Language (DQL) commands. These two types of SQL commands are especially useful:

- Data Manipulation Language (DML):

    1. **Insight extraction**: DML commands are used to insert, update, delete, and manage data within database tables. The primary goal of data analysis is to extract insights from data rather than modify it.

    2. **Data preparation**: Before analyzing data, it often needs to be cleaned and preprocessed. DML commands like UPDATE can correct data errors, and DELETE can remove irrelevant or duplicate records. Data preparation is crucial for accurate analysis.

3. **Inserting data**: The INSERT command is useful for adding new data to the database, which might be needed for analysis. This could include new data points, calculated metrics, or results from previous analyses that you want to store for future use.

- Data Query Language (DQL):

    1. **Data retrieval**: The essence of data analysis in SQL environments is to query the database for specific datasets. SELECT allows you to specify exactly which data to retrieve, including which tables to source from and under what conditions.

    2. **Data aggregation and filtering**: SELECT queries can be augmented with clauses like WHERE, GROUP BY, and HAVING. They filter data, aggregate it (e.g., finding averages, sums, counts), and select data that meets certain conditions. These operations are fundamental to data analysis, enabling analysts to explore trends, patterns, and outliers in the data.

    3. **Joining tables**: Data analysis often requires combining data from multiple tables to get a complete picture. The SELECT command can join tables based on specific criteria, enabling comprehensive analysis across diverse datasets.

- Both DML and DQL:

    1. **Flexibility in data handling**: DML provides the flexibility to manipulate data as needed for analysis, ensuring the dataset is accurate and relevant. DQL offers the tools to dig into that data, pulling out the insights and information critical to informed decision-making.

    2. **Basis for advanced analysis**: While other SQL commands focus on database structure and access control, DML and DQL are directly concerned with the data itself. Mastery of these commands allows analysts to extract and manipulate data.

    3. **Data integrity**: While DML helps maintain the quality and relevance of the dataset, DQL ensures that the integrity of the data is preserved during analysis. By using DQL, analysts can perform read-only operations that don't risk altering or damaging the underlying data.

DML and DQL are essential for data analysis. DML prepares the data landscape for analysis, ensuring that it is clean and current, whereas DQL enables analysts to query, aggregate, and interpret the data to derive actionable insights from it. Together, they form the backbone of data analysis in SQL databases.

Transaction Statements vs Query Statements in Data Analysis

The distinction between transaction statements and query statements in SQL refers to two fundamental aspects of data analysis: managing the integrity of data operations and gaining insights.

Transaction Statements in Data Analysis

Transaction statements, which manage how data changes are applied or reverted, play a crucial role in ensuring the integrity and consistency of the data throughout the analysis process. When multiple operations are performed, transaction statements allow analysts to maintain consistent data states. For instance, updating a database to correct errors or reflect new information in a transaction ensures that all updates are applied successfully, or none are applied at all. This prevents partial updates that could cause data inconsistencies. Data analysts often need to experiment with data transformations or corrections. Transactions provide safety (using BEGIN, COMMIT, and ROLLBACK), which allows analysts to test changes without permanently altering the data until they are certain of the results.

Query Statements in Data Analysis

Query statements, focused on data retrieval, are the foundation of data analysis. They enable analysts to explore, aggregate, and visualize data, extracting meaningful insights. It is the process of exploring datasets in order to gain a better understanding of underlying patterns, trends, and anomalies that constitutes the heart of data analysis. Query statements (SELECT) allow analysts to sift through large volumes of data, filter specific subsets, and perform complex joins across tables to gather comprehensive insights.

Integrating Transaction and Query Statements in Data Analysis

While transaction statements provide a safe means of manipulating data, query statements provide a means of extracting insights. In an organized data analysis workflow, analysts may use transaction statements to prepare data for analysis (e.g., correcting, updating, or cleaning data), and query statements to extract insights from the data once it has reached a reliable state. It should be noted that transactions can also play a role in ensuring that the data manipulation steps are reproducible and reversible, which is critical for verifying and validating the analysis results.

Setting Up a Storytelling Environment with PostgreSQL

Creating a storytelling environment with PostgreSQL involves setting up a database system where data can be stored, manipulated, and queried to uncover and narrate compelling stories hidden within data. PostgreSQL offers an ideal platform for data analysis and storytelling. This section guides you through the initial steps to establish such an environment, emphasizing the simplicity and power of PostgreSQL.

Step 1: Installation

Download and install PostgreSQL. Download the PostgreSQL installer for your operating system from the official PostgreSQL website. The installation process is straightforward. Follow the on-screen instructions and make sure to note the administrator password you set during installation as well as the default port on which PostgreSQL will run (usually 5432).

Step 2: Create Your First Database

Access PostgreSQL. After installation, access PostgreSQL through its command-line interface (CLI) known as `psql`, or use a graphical user interface (GUI) tool. Both provide comprehensive database management capabilities, with `pgAdmin` being more beginner-friendly due to its visual nature.

Create a database. To create your first database, use the `psql` command-line interface or pgAdmin. In PostgreSQL, you can create a database named `storytelling_db` by executing this command:

```
CREATE DATABASE storytelling_db;
```

Step 3: Define Data Structures

Create the tables. With your database in place, the next step is to define the structure of your data by creating tables. For example, if you're telling stories about customer interactions, you might create a table called `customers` with fields for customer ID, name, email, and interaction dates.

```
CREATE TABLE customers (
    customer_id SERIAL PRIMARY KEY,
    name VARCHAR(100),
    email VARCHAR(100),
    interaction_date DATE
);
```

Insert the data. Fill your table with initial data to start your analysis. INSERT statements add records to your tables, laying the groundwork for your storytelling.

```
INSERT INTO customers (name, email, interaction_date) VALUES
('Jane Doe', 'jane.doe@email.com', '2022-01-01'),
('John Smith', 'john.smith@email.com', '2022-01-02');
```

The database named `storytelling_db`, contains a table called `customers`. Table 1-2 illustrates the `customers` table, which was created to hold customer interaction data. This table is structured to capture essential details about customers and their interactions. The fields within the customers table are as follows:

- `name`: The customer's name, stored as a variable character string (VARCHAR) with a maximum length of 100 characters.

- `email`: The customer's email address, also stored as a VARCHAR with a maximum length of 100 characters.

- `interaction_date`: The date of the interaction with the customer, stored as a DATE.

This structure allows you to store and query customer interaction data effectively. The next section discusses the different data types of variables in SQL.

Table 1-2. *The Customer Table*

customer_id	Name	Email	interaction_date
1	Jane Doe	jane.doe@email.com	2022-01-01
2	John Smith	john.smith@email.com	2022-01-02

Setting up an environment using PostgreSQL is a crucial step. The initial steps in creating a robust analysis and storytelling platform are installation, database creation, data structuring, and querying. In the forthcoming chapters of this book, you are taken on a detailed, step-by-step journey into the art of query writing. Through the examples provided in the next chapters, you will develop SQL skills, from the basics to advanced querying techniques.

Data Types in SQL

Data types in SQL are an essential concept, defining the nature of data that can be stored in a database column. Each data type in SQL ensures that data fits into predefined formats, facilitating accurate data storage, retrieval, and analysis. To design and manipulate databases effectively, it is crucial to be familiar with these data types. The following are common data types available in SQL:

- Numeric data types:

    INTEGER: A whole number, either positive or negative. Depending on the database system, variations like INT, SMALLINT, TINYINT, and BIGINT represent integers of different sizes.

    DECIMAL and NUMERIC: These data types are designed to store exact numeric data values. Defining DECIMAL or NUMERIC columns in SQL allows analysts to specify precision and scale so that the database handles numeric data precisely. Precision refers to the amount of significant digits in a number, to both left and right of the decimal point. It's the total count of digits in the number. The scale specifies the number of digits after the decimal point. It represents the fraction part of the number and is a subset of the precision.

FLOAT, REAL, and DOUBLE PRECISION: Represent floating-point numbers with varying levels of precision. Ideal for scientific calculations where exact precision is not critical.

- String data types:

CHAR and CHARACTER: A fixed-length string. If the entered string is shorter than the specified length, it will be right-padded with spaces.

VARCHAR and CHARACTER VARYING: Variable-length strings. Allows for storing strings up to a specified maximum length.

TEXT: For large text data where the length might exceed the limits of VARCHAR.

- Date and time data types:

DATE: Stores date values, including year, month, and day.

TIME: Stores time of day values.

TIMESTAMP: Combines date and time, capturing a specific moment in time.

INTERVAL: Represents a span of time, useful for calculating differences between dates or times.

- Boolean data type:

BOOLEAN: Represents logical Boolean values, typically as TRUE or FALSE.

- Binary data types:

BINARY and VARBINARY: Stores binary data, such as images or files, in fixed-length or variable-length formats, respectively.

BLOB (Binary Large Object): For storing large binary data, like images, videos, or documents.

- Specialized data types:

ENUM: A string object that can have only one value chosen from a list of values defined at the table creation time.

ARRAY: Supports storing an array, which is an ordered collection of elements, of a specified data type.

JSON and XML: For storing JSON or XML data, allowing for complex data structures within a single database column.

UUID: Stores Universally Unique Identifiers.

- Geospatial data types:

POINT, LINESTRING, and POLYGON: Specific to databases that support geospatial data for representing geographic shapes and locations.

It is important to choose the appropriate data type to optimize database storage, performance, and data integrity. Consider your data characteristics and performance requirements when designing your database schema.

The following are examples of each of the mentioned SQL data types, and they illustrate how these data types can be used when creating a table.

- Numeric data types:

```
CREATE TABLE numeric_examples (
    id INT,
    small_number SMALLINT,
    big_number BIGINT,
    exact_amount DECIMAL(10, 2),
    approx_net_worth FLOAT
);
```

This query creates a table named numeric_examples with five columns, each designated to hold numeric data of various types and scales. The columns include an integer called id, a small integer called small_number, a large integer called big_number, a precise decimal called exact_amount suitable for financial data, and a floating-point number called approx_net_worth for approximate values.

- String data types:

```
CREATE TABLE string_examples (
    fixed_char CHAR(10),
    variable_char VARCHAR(100),
    long_text TEXT
);
```

15

This query creates a table named `string_examples` composed of three columns designed to store string data in distinct formats. The `fixed_char` column stores fixed-length strings of ten characters; `variable_char` accommodates variable-length strings up to 100 characters; and `long_text` stores large text entries without a specified maximum length.

- Date and time data types:

```
CREATE TABLE datetime_examples (
    birth_date DATE,
    appointment_time TIME,
    event_timestamp TIMESTAMP,
    duration INTERVAL
);
```

This query creates a table called `datetime_examples`, which is designed to store various types of date and time information across four columns. The table includes a `birth_date` column for dates, an `appointment_time` column for times of the day, an `event_timestamp` for date and time combinations, and a `duration` column to represent time intervals.

- Boolean data type:

```
CREATE TABLE boolean_example (
    is_active BOOLEAN
);
```

This query generates a table named `boolean_example` that consists of a single column, called `is_active`. The `is_active` column is designed to store Boolean values, indicating a true or false state.

- Binary data types:

```
CREATE TABLE binary_examples (
    fixed_binary BYTEA,
    variable_binary BYTEA,
    large_object OID
);
```

This query constructs a table named binary_examples with three columns designed to store binary data in various formats. It includes a fixed_binary column for storing binary data using the BYTEA type, which handles variable-length binary data. The variable_binary column is also of type BYTEA, allowing flexible storage of binary data without a predefined size limit. Finally, the large_object column is used to store large binary objects (BLOBs) using the OID type, which refers to large objects stored separately in PostgreSQL. With this setup, different types of binary data can be managed efficiently in a robust way.

- Specialized data types:

```
CREATE TYPE status_enum AS ENUM ('New', 'In Progress',
'Completed');

CREATE TABLE specialized_examples (
    status status_enum,
    number_series INTEGER[],
    user_profile JSON,
    unique_id UUID
);
```

In PostgreSQL, to define the ENUM type, you have to use CREATE TYPE. In this query, the CREATE TYPE statement defines an ENUM type, called status_enum, which can then be used in table definitions.

This query establishes a table called specialized_examples, incorporating columns with specialized data types: status as an ENUM to restrict values to specific states, number_series as an ARRAY to store sequences of integers, user_profile for storing structured JSON data, and unique_id to hold universally unique identifiers (UUIDs).

Note   The `--` symbol in SQL signifies the start of a single-line comment, indicating that the text following it on the same line is not executed as part of the SQL command and is used for annotations or explanations within the script.

In SQL, the semicolon (`;`) is a syntax element that serves several purposes, including like statement terminator, batch processing, compatibility, and clarity. In PostgreSQL, the use of semicolons is more strict compared to some other database management systems like MySQL. In PostgreSQL's interactive terminal (`psql`), semicolons are generally required to terminate SQL statements. Without a semicolon, the terminal waits for further input, assuming the statement is not yet complete. In PostgreSQL, semicolons are required in interactive sessions (`psql`) to execute statements, mandatory to separate multiple statements, necessary in scripts to ensure each statement is processed correctly, and used in PL/pgSQL code blocks to terminate individual statements.

Note   Geospatial types are specific to databases that support them, like PostgreSQL with PostGIS.

- Geospatial data types:

```
CREATE TABLE geospatial_examples (
    location_point POINT,
    route LINESTRING,
    boundary POLYGON
);
```

This query creates a table named `geospatial_examples` in a database system that supports geospatial data types, such as PostgreSQL with the PostGIS extension, designed to store various types of geographical data. The table includes three columns: `location_point` for storing a single geographical point, `route` for a series of connected points forming a line, and `boundary` for defining a closed shape or area.

Crafting the Narrative

Throughout the remainder of this book, each chapter will examine SQL operators using a unique teaching methodology that combines data analysis and storytelling. By exploring various SQL operators and their applications, each chapter will uncover insights and create compelling narratives from data. By enhancing your technical skills as well as your analytical thinking, you will be able to communicate complex data-driven stories effectively.

As the name of this book indicates, *narrative SQL* refers to the process of translating natural language queries into SQL commands. A natural language query refers to a question or command expressed in everyday language that humans use to communicate, rather than in a specialized programming or query language. This enables users to interact with systems, databases, and computers by using human-like sentences. In this book, you are invited on a journey through the world of data querying and manipulation, where the complexities of SQL are unraveled in a narrative manner. Following the adventures of a detective solving mysteries using SQL queries based on data provided from a database. Each chapter is structured as a series of narratives, making technical content approachable and memorable. Through the use of compelling stories, this book will not only educate, but also inspire readers from all backgrounds to learn SQL concepts and use them with confidence in their data-driven endeavors.

Summary

This chapter began with the fundamentals of data and its storage in databases to provide the foundations for mastering SQL through data storytelling. It explored the distinctions between relational and non-relational databases and discussed why PostgreSQL is particularly suited to learning SQL, with its robust features and strong community support. Each section builds on the last, from understanding the various SQL commands—categorized into DDL, DML, DCL, TCL, and DQL—to setting up a PostgreSQL environment conducive to storytelling.

Key Points

- **Data**: Data consists of a wide range of information types that can be digitally stored, processed, and analyzed, serving as the foundation for insights and decision-making.

- **Databases**: Databases can play a fundamental role in storing, organizing, and managing data, setting the stage for effective data analysis and storytelling.

- **Relational vs. non-relational databases**: Relational databases organize data into tables linked by relationships, whereas non-relational databases store data in a non-tabular form, in response to diverse data storage needs.

- **Choosing PostgreSQL**: PostgreSQL offers robustness, a wide range of features, as well as strong community support, making it a suitable choice for learning and using SQL in data storytelling.

- **Understanding SQL**: SQL is an essential language for database interaction, covering its significance in querying, updating, and managing data.

- **SQL command categories**: SQL commands are categorized into DDL, DML, DCL, TCL, and DQL, providing a foundational understanding of their roles in database operations.

- **Setting up a storytelling environment**: Setting up a PostgreSQL environment, from installation to creating their first database and tables, enables data-driven storytelling.

- **SQL data types**: SQL data types (numeric, string, date and time, Boolean, binary, and specialized types) enable accurate storing and manipulating of data.

- **Getting started with data analysis and storytelling**: Through SQL queries, a journey of uncovering patterns, trends, and insights begins to lay the foundation for narratives.

Key Takeaways

- **Data and databases**: Essential for storing and analyzing information, which serves as the backbone for decision-making.

- **PostgreSQL**: Chosen for its robustness and suitability for educational purposes in SQL and data storytelling.

- **SQL basics**: Covered the importance of SQL in database operations and introduced the basic commands and data types.

As you've explored SQL, you've learned that it can be used not just to query data. SQL can also tell compelling stories. After setting up the storytelling environment, you are now ready to dive deeper into SQL's complexities.

Looking Ahead

As you move into the next chapter, "Starting with SELECT," you will delve into how to retrieve and manipulate data effectively using SQL's most frequently used command. This involves learning to craft precise SELECT statements to extract just the right data needed for your narratives. This will set the stage for more advanced data manipulation and analysis techniques.

CHAPTER 2

Starting with SELECT

In this chapter, you learn about the SQL command that is fundamental for retrieving data from databases. This chapter demonstrates the versatility of the SELECT statement by using narratives and practical examples to demonstrate how it is used in both simple and complex data retrieval scenarios.

Introduction to SELECT

The SELECT statement is one of the most fundamental and frequently used SQL commands, standing at the heart of almost every query operation. For any SQL user, it is an essential tool for retrieving data from databases. At its core, the SELECT statement allows you to specify exactly what data you want to retrieve from which tables, and it can be fine-tuned with various clauses to meet precise data retrieval needs. SELECT is not just about pulling data from a database; it is about choosing the right pieces of data that answer specific questions or fulfill particular informational needs. As a result, the SELECT statement can transform raw data into meaningful information.

The Importance of SELECT in Storytelling with Data

When it comes to data storytelling, SELECT goes beyond data retrieval to become a tool that assists in narrative construction. Data storytelling is about weaving data into a narrative that makes sense to the audience, helping them understand complex information through real-world stories.

For instance, to extract the data points that form the basis of these stories, SELECT can be used to extract a wide variety of data points. Consider a dataset that contains years of sales data that was collected over a period of time. The SELECT command could be used by a storyteller to retrieve total sales during major events or holidays and analyze them. Careful selection of data can reveal more about consumer behavior during certain

© Hamed Tabrizchi 2025
H. Tabrizchi, *Narrative SQL*, https://doi.org/10.1007/979-8-8688-1560-7_2

periods. As a result of selecting specific data, storytellers can emphasize information that illustrates trends, supports a hypothesis, or explains phenomena in a way that is visually and contextually compelling.

The Anatomy of a SELECT Statement

A SELECT statement's structure and syntax are crucial to maximizing SQL's power in data retrieval and storytelling. This subsection breaks down the fundamental components of the SELECT statement, including how to select columns and use aliases to make queries more readable.

Basic Structure and Syntax

The SELECT statement is used to query the database and retrieve specified data. At its most basic, a SELECT statement must specify two key pieces of information: what you want to select and from where. The syntax is easy to understand:

```
SELECT column1, column2, ...
FROM tableName;
```

In this structure:

- SELECT indicates that you are about to specify a list of columns or expressions you want to retrieve.

- column1, column2, ... are the specific columns you want to select from the database table. You can also use * character to select all columns from a table.

- FROM tableName specifies the table from which to retrieve the data.

This simple syntax is the starting point for building more complex queries, including those that filter, group, or sort data.

The First Story: The Bookstore Anniversary

This story is based on a database named Customers. Each of the characters has a unique background and story to tell. John Doe, Jane Smith, and Alice Johnson are regular clients of a bookstore. The bookstore is celebrating its anniversary next month. This event will be a success if its valued customers attend. This requires a personalized invitation for each of registered customers, and the key information needed for these invitations includes each customer's full name and email address. Table 2-1 shows the table named Customers.

Table 2-1. *The Customers Table*

CustomerID	FirstName	LastName	Email	DateOfBirth	Neighborhood	Address	ZIP Code	Phone Number
1	John	Doe	johndoe@example.com	1980-05-15	Downtown	123 Elm St.	90210	(555) 321-9876
2	Jane	Smith	janesmith@example.com	1975-07-20	Midtown	456 Oak Ave.	90212	(555) 654-1234
3	Alice	Johnson	alicej@example.com	1990-11-12	Eastside	789 Pine Rd.	90213	(555) 789-2342

Selecting Columns from a Table

When constructing a narrative or conducting an analysis, you often need specific pieces of data from one or more tables. Selecting specific columns allows you to focus on the data relevant to your story or analysis. For instance, if you have a database table named Customers that includes CustomerID, FirstName, LastName, Email, DateOfBirth, Neighborhood, Address, Zip Code, and Phone Number, and your story or analysis only needs to know about customer names and emails, the proper SELECT statement would look something like this:

```
SELECT FirstName, LastName, Email
FROM Customers;
```

Table 2-2 shows the result of executing the query to select only the FirstName, LastName, and Email columns.

Table 2-2. *The Result of the SELECT Query Execution*

FirstName	LastName	Email
John	Doe	johndoe@example.com
Jane	Smith	janesmith@example.com
Alice	Johnson	alicej@example.com

Introducing Aliases for Columns

Aliases in SQL are used to rename a column or table in the output of your SQL query. They are particularly useful for making query results more readable. For example, using Contact Email instead of just Email can significantly enhance clarity. The same applies when joining multiple tables, which is covered in Chapter 4. It is also helpful to use aliases when joining multiple tables with columns with the same name but with different meanings. The AS keyword is used after the column name to define an alias. However, the AS keyword is optional.

```
SELECT FirstName AS First, LastName AS Last, Email AS ContactEmail
FROM Customers;
```

In this query, `FirstName`, `LastName`, and `Email` are renamed to `First`, `Last`, and `ContactEmail`, respectively, in the output. Aliases are especially useful in reports and exported data, making them essential tools in data presentation and storytelling. See Table 2-3.

Table 2-3. *The Result of the SELECT Along with AS Query Execution*

First	Last	ContactEmail
John	Doe	johndoe@example.com
Jane	Smith	janesmith@example.com
Alice	Johnson	alicej@example.com

Introducing the CONCAT Function

To complete the process, the final step is to construct the following SQL query, which concatenates the first and last names to retrieve the full name and email address.

```
SELECT CONCAT(FirstName, ' ', LastName) AS FullName, Email
FROM Customers;
```

This SQL statement utilizes the `CONCAT()` function to merge the `FirstName` and `LastName` into a single `FullName` column, simplifying the presentation and personalization of each invitation. By selecting only the `FullName` and `Email` columns, this query efficiently pulls the essential information, ensuring that the process of creating and sending invitations is streamlined and targeted. See Table 2-4.

Table 2-4. *The Result of the Query Execution of the SELECT, AS, and CONCAT Functions*

FullName	Email
John Doe	johndoe@example.com
Jane Smith	janesmith@example.com
Alice Johnson	alicej@example.com

This table efficiently presents the necessary information for sending the bookstore anniversary invitations, presenting only the combined full names and email addresses of each customer.

Note   CONCAT is a function in SQL, not an operation. It's used to concatenate, or join together, two or more strings into a single string. The CONCAT function is an essential tool for combining data from different columns or for creating a formatted output from multiple string fields in a database. The basic syntax of the CONCAT function is as follows:

```
CONCAT(string_column_1, string_column_2, ..., string_cloumn_N)
```

It should be noted that the CONCAT function in PostgreSQL can be used in a variety of scenarios. However, there are specific cases where it may not be applicable or where alternative methods might be more appropriate. The || operator is a standard way to concatenate strings in PostgreSQL and can serve as an alternative to the CONCAT function. For instance, consider the following example using the CONCAT function:

```
SELECT CONCAT(string_column_1, string_column_2, ..., string_column_N) AS concatenated_string

FROMyour_table;
```

Alternatively, it is possible to achieve the same result using the || operator, which is a standard SQL method for string concatenation:

```
SELECT string_column_1 || string_column_2 || ... || string_column_N AS concatenated_string

FROM your_table;
```

The || operator is widely supported across various SQL databases, making it a more portable option compared to some proprietary functions like CONCAT. Both approaches effectively concatenate strings, and the choice between them is influenced by factors such as readability, performance considerations, and specific database requirements.

SQL Mathematical Operations with SELECT

In SQL, mathematical operations are crucial to transforming and analyzing data. These operations enable you to perform calculations across your data columns, enhancing the ability to communicate meaningful stories through numbers. This section explores the basic mathematical operations available in SQL, illustrates these operations with a summary table, and shows how these techniques can be applied to storytelling.

SQL supports a variety of mathematical operations that can be applied directly in the SELECT statement. These include addition (+), subtraction (-), multiplication (*), and division (/). Data manipulation can be performed directly within the database query using these operations to compute new values from existing data. This query provides a basic example of using mathematical operations to calculate a simple product:

```
SELECT column1, column1 * column2 as Product
FROM Table;
```

In this query, column1 and column2 are existing columns in Table, and Product is a new column created in the output that contains the product of column1 and column2.

Table 2-5 summarizes some of the most commonly used SQL mathematical operations.

Table 2-5. *An Illustration of the Most Useful SQL Mathematical Operations*

Operation	SQL Symbol	Example Use	Description
Addition	+	column1 + column2	Adds two columns together.
Subtraction	-	column1 - column2	Subtracts one column from another.
Multiplication	*	column1 * column2	Multiplies two columns.
Division	/	column1 / column2	Divides one column by another.
Modulus	%	column1 % column2	Divides one column by another and returns the remainder.

These operations can be used to create derived columns and provide greater insight into your data.

The Second Story: The Bakery Sales Data

There is a small bakery that wants to analyze the sales data it has collected. They have a table named Sales with columns like CustomerID, ProductName, Price, RegularPrice, Discount, and Quantity. They raise the following questions in their analysis:

- Which flavor of donut is most popular?

- Who are our most frequent customers?

- How much revenue does each product generate?

Before answering each question, consider the Sales table, shown in Table 2-6.

Table 2-6. *The Sales Table*

CustomerID	ProductName	Price	Quantity	RegularPrice	Discount
743663	Glazed Donut	2.5	2	3.0	0.5
743663	Chocolate Muffin	3.0	1	3.0	0.0
223424	Blueberry Muffin	3.0	2	3.0	0.0
323423	Chocolate Chip Cookie	1.5	3	2.0	0.5
432424	Sugar Cookie	1.0	4	1.0	0.0

The first question can be answered with the following query. They can use a simple SELECT statement in order to determine the most popular donut flavor.

```
SELECT ProductName, SUM(Quantity) AS TotalSold
FROM Sales
GROUP BY ProductName;
```

This query selects ProductName and calculates the total quantity sold using SUM(Quantity) with an alias TotalSold. SUM() is a function in SQL used to calculate the total of a numeric column in a table. It should be noted that this function calculates the sum of all values in a specified numeric column and ignores NULL values by default. In this query, GROUP BY ProductName groups all sales records by product name, so the SUM(Quantity) calculates the total quantity sold for each unique product. This helps summarize data per product instead of across the entire table. See Table 2-7.

Table 2-7. *The Result of First Question Query Execution*

ProductName	TotalSold
Chocolate Chip Cookie	3
Glazed Donut	2
Blueberry Muffin	2
Chocolate Muffin	1
Sugar Cookie	4

Note   In SQL, the GROUP BY clause is used to group rows that have the same values in specified columns into summary rows. It's commonly used with aggregate functions like SUM(), AVG(), COUNT(), and MAX() to perform calculations on each group of data. Chapter 5 covers the GROUP BY clause in detail, while Chapters 5 and 10 cover a much broader range of aggregate functions.

To answer the second question (most frequent customers), you could use the following query:

```
SELECT CustomerID, COUNT(*) AS PurchaseCount
FROM Sales
```

This query selects Customer ID and counts the number of purchases using COUNT(*) with an alias PurchaseCount. COUNT(*) in SQL is a function used to get the total number of rows in a table. It should be noted that this function counts all rows in a table, including duplicates and rows with NULL values.

Note   COUNT(*) is distinct from COUNT(column_name), which counts only non-NULL values in a specific column. In SQL, the asterisk (*) is used to select all columns. The asterisk represents all columns from a table. Using * in SELECT can be inefficient for large tables, as it retrieves all the data. Consider selecting specific columns for better performance.

Table 2-8 shows the result of executing the query to select only the `CustomerID` and `PurchaseCount` columns.

Table 2-8. *The Result of the Second Question Query Execution*

CustomerID	PurchaseCount
743663	2
223424	1
323423	1
432424	1

To answer the third question (how much revenue does each product generate), the following query can be used:

```
SELECT
  CASE
    WHEN ProductName LIKE '%Donut%' THEN 'Donuts'
    WHEN ProductName LIKE '%Muffin%' THEN 'Muffins'
    ELSE 'Cookies'
  END AS Category,
  SUM(Price * Quantity) AS TotalRevenue
FROM Sales
GROUP BY
  CASE
    WHEN ProductName LIKE '%Donut%' THEN 'Donuts'
    WHEN ProductName LIKE '%Muffin%' THEN 'Muffins'
    ELSE 'Cookies'
  END;
```

This query is a bit more complex. It uses a `CASE` statement to categorize products based on their names. It calculates total revenue using `SUM(Price * Quantity)`. In order to find the answer, the query analyzes product names and categorizes them into `Donuts`, `Muffins`, or `Cookies`. Then, it multiplies the price of each item by the quantity sold and adds them to get the total revenue generated by each category. In this way, the revenue generated by each product category can be calculated.

It should be noted that the % symbol is a special character used to perform pattern matching within strings. % represents any sequence of characters, zero or more characters, within a pattern. For instance, %Donut% is a matching pattern. In this specific case, the query is looking for any product name that contains the substring "Donut" anywhere within it. So, it would match product names like "Glazed Donut" and "Chocolate Donut with Sprinkles".

Finally, the GROUP BY clause is used with the same CASE expression to group the records by their assigned category (Donuts, Muffins, or Cookies) before applying the SUM function. This ensures correct revenue aggregates for each product category. Table 2-9 shows the result.

Table 2-9. *The Result of Third Question Query Execution*

Category	TotalRevenue
Donuts	5
Muffins	9
Cookies	8.5

As shown in Table 2-9, the CASE statement categorized the products. Products with "Donut" in their name became Donuts, "Muffin" became Muffins, and everything else became Cookies. SUM(Price * Quantity) calculated the total revenue for each category.

The CASE statement

The CASE statement in SQL acts like if-then-else logic for queries. It evaluates a series of conditions and returns a corresponding value based on the first matching condition.

```
CASE
  WHEN condition1 THEN result1
  WHEN condition2 THEN result2
  ...
  ELSE result_else
END
```

- The WHEN statement specifies a condition to be evaluated. It can be any valid SQL expression that returns a Boolean value (true/false).

- If the corresponding WHEN condition evaluates to true, this value is returned by THEN.

- ELSE is an optional clause that provides a default result if none of the WHEN conditions are met. If omitted and no condition is true, it usually returns NULL.

For more complex queries, it is possible to nest CASE statements. You learn more about CASE and nested CASEs in Chapter 8 in particular.

String Patterns

A string pattern in SQL allows you to search and filter data based on specific patterns within text columns. There are two common wildcards used in string patterns in SQL: the % symbol, which represents any number of characters, and the _ symbol, which represents a single character. These wildcards provide flexibility in searching and filtering data based on varying patterns. Chapter 3 explores more complex examples of string matching after you learn about the WHERE and LIKE statements.

Following the answers to the three questions, Sarah is in the process of preparing for a promotional event. Her goal is to offer discounts on some items, but she is unsure which ones would have the greatest impact on her customers. Ideally, she wants to target items that are:

- **Popular**: Items with a high number of sales.

- **Not already discounted**: She does not want to discount already discounted items.

To identify these ideal candidates for promotion, Sarah decides to analyze her sales data. Using these two CASE statements, the following query can achieve this.

```
SELECT ProductName, Quantity,
  CASE
    WHEN Quantity >= (SELECT AVG(Quantity) FROM Sales) THEN 'Popular'
    ELSE 'Less Popular'
```

```
END AS Popularity,
CASE WHEN RegularPrice > Price THEN 'Discounted' ELSE 'Regular Price' END
AS Price_Status
FROM Sales
```

In the first CASE, the average quantity is calculated using a subquery, and the item is categorized based on whether it exceeds or equals the average quantity. The second CASE statement checks the price status. This ensures that Sarah focuses on discounts on popular items. If the item is Popular (based on the outer CASE) and the regular price (RegularPrice) is greater than the current price (Price), it means the item is discounted. If the item is Popular but doesn't have a discount (regular price equals current price), it's categorized as Regular Price.

Note   Subqueries, or inner queries or nested queries, are powerful SQL tools for embedding SELECT statements within each other. They are used to retrieve data for the outer query. This type of queries are discussed in more detail in the next chapters, as you become more aware of their importance.

Table 2-10 is the result of executing the query to select only the ProductName, Quantity, calculated Popularity and PriceStatus columns.

Table 2-10. *The Result of Promotional Event Query Execution*

ProductName	Quantity	Popularity	PriceStatus
Glazed Donut	2	Less Popular	Discounted
Chocolate Muffin	1	Less Popular	Regular Price
Blueberry Muffin	2	Less Popular	Regular Price
Chocolate Chip Cookie	3	Popular	Discounted
Sugar Cookie	4	Popular	Regular Price

The Art of Distinct Selection

To create compelling narratives through data, it is essential to utilize the `DISTINCT` keyword effectively. In SQL, the `DISTINCT` keyword refines data retrieval by ensuring that query results contain unique values. It acts like a filter, eliminating duplicate rows that might skew your analysis or storytelling. Here's the basic syntax:

```
SELECT DISTINCT column
FROM table
```

This retrieves only unique values in the specified `column` from the `table`.

`DISTINCT` removes rows where all selected columns have identical values. For example, if a table has `CustomerID` and `OrderDate`, `DISTINCT CustomerID` will return each unique customer, even if they placed multiple orders.

The `DISTINCT` statement can ensure uniqueness across multiple columns when used with a combination of columns. For instance, `SELECT DISTINCT CustomerID, ProductID` will return only rows where the customer and product combinations are both distinct.

You can also combine `DISTINCT` with aggregation functions such as `SUM` to uncover trends within your data.

The Third Story: The Candy Store Sales Data

Phoebe is eager to extend her vibrant candy store overflowing with colorful treats. Her sales report includes today's purchases, and she is curious about her customers' favorites. Table 2-11 shows the sales log.

Table 2-11. *The Sales Log*

OrderID	Candy	Quantity
1	Gummy Bears	2
2	Lollipops	1
3	Gummy Bears	3
4	Chocolate Bars	2
5	Lollipops	2
6	Gummy Bears	1

The log shows every sale, but not the unique candies her customers crave. The magnifying glass she needs is DISTINCT:

```
SELECT DISTINCT Candy
FROM sales_log
```

This query uncovers the data in Table 2-12.

Table 2-12. *Unique Candies from the Sales Log Table*

Candy
Gummy Bears
Lollipops
Chocolate Bars

Note   Using DISTINCT on large datasets can impact performance. It is important to consider whether you really need all possible combinations or if filtering beforehand is sufficient. The upcoming chapters discuss data filtering in more detail.

Aggregating with SELECT

In SQL, aggregate functions are essential for calculating characteristics of a dataset and producing a single value that summarizes their characteristics. These functions allow analysts to extract more meaning from the data, transforming raw data into useful insights that support decision-making and storytelling. Data analysis in SQL depends heavily on aggregate functions, which allow computations across large volumes of data in order to create efficient summaries and high-level overviews.

Differences Between Regular Arithmetic Functions and Aggregate Functions in SQL

Regular arithmetic functions in SQL are useful for manipulating data within individual rows. They perform calculations like addition, subtraction, multiplication, and more, transforming or combining values on a row-by-row basis. Aggregate functions, on the other hand, take a broader view. Summarizing entire groups of data is one of their strengths. The COUNT, SUM, AVG, MIN, and MAX functions combine multiple rows into a single, meaningful value, revealing total quantities, averages, and high and low values within a specific column. In contrast to regular arithmetic functions, aggregate functions help you identify trends and patterns across your data.

The following lists describe some of the key differences between regular arithmetic functions and aggregate functions in SQL.

Regular arithmetic functions:

- **Purpose**: Perform calculations on a row-by-row basis. They manipulate data within a single row or combine values from multiple columns within a row.

- **Examples**: +, -, *, /, MOD (modulo), ROUND, SQRT, ABS (absolute value), and so on.

- **Output**: They return a single value for each row processed in the query.

- **Focus on individual values**: They operate on individual values within a row, transforming or combining them as needed.

Aggregate functions:

- **Purpose**: Summarize data by performing calculations on entire groups of values. They condense multiple rows into a single, meaningful value.

- **Examples**: SUM, COUNT, AVG, MIN, and MAX.

- **Output**: They return a single value that represents the overall result of the calculation across a group of rows.

- **Focus on summarization**: They focus on providing a summary statistic (total, average, minimum, maximum, etc.) for the data in a specific column or set of columns.

Table 2-13 summarizes the differences between regular arithmetic functions and aggregate functions.

Table 2-13. *Differences Between Regular Arithmetic Functions and Aggregate Functions*

Feature	Regular Arithmetic Functions	Aggregate Functions
Purpose	Row-by-row calculations	Summarize data groups
Input	Values within a row	Groups of rows
Output	Single value per row	Single value
Focus	Individual values manipulation	Summarization

As an example, to illustrate the difference between these two, consider the Orders table shown in Table 2-14, with the OrderID, CustomerID, and Amount columns.

Table 2-14. *The Orders Table*

OrderID	CustomerID	Amount
1	101	100
2	102	50
3	103	75
4	101	275

This query calculates a 10 percent discount for each order row-by-row multiplication, as shown in Table 2-15:

```
SELECT OrderID, CustomerID, Amount, Amount * 0.1 AS Discount
FROM Orders;
```

Table 2-15. *Discount Added to the Orders Table*

OrderID	CustomerID	Amount	Discount
1	101	100	10
2	102	50	5
3	103	75	7.5
4	101	275	27.500

This query calculates a 10 percent discount for each order by multiplying the Amount by 0.1 and adds a new column named Discount to the result set.

The following query uses aggregate functions to get the total number of orders (COUNT(*)) and the total sales amount (SUM(Amount)) for all orders.

```
SELECT
  CustomerID,
  COUNT(*) AS TotalOrders,
  SUM(Amount) AS TotalSales
FROM Orders
GROUP BY CustomerID;
```

As shown in Table 2-16, this query uses aggregate functions to get the total number of orders and the total sales amount for all orders.

Table 2-16. *The Total Number of Orders and Total Sales Amount for All Orders*

Customerid	TotalOrders	TotalSales
101	2	375.00
102	1	50.00
103	1	75.00

Table 2-17 illustrates a number of primary aggregate functions that play pivotal roles in data analysis in SQL.

Table 2-17. *A Number of Primary Aggregate Functions*

Aggregate Function	Description
COUNT	Used to count the number of items in a particular column or dataset, helping determine the size, extent, or frequency of various categories within the data. For example, COUNT can tell how many books are in each category of a bookstore's inventory, providing a quantitative base for inventory management decisions.
AVG	Used to calculate the average value of a numeric column, which is crucial for understanding typical values when dealing with variables such as prices, ages, or any measurable quantities where averages can provide insights into normal behavior or expected outcomes. For instance, determining the average number of pages in their books can help publishers understand typical publication lengths within genres.
MAX	Finds the highest value in a column. This is particularly useful when you need to identify peaks or maximum levels in datasets, such as finding the most expensive item sold in a store or the highest score achieved by an individual. MAX can highlight outliers or exceptional cases in data analysis.
MIN	Conversely, the MIN function determines the lowest value in a column. It can be crucial for identifying the least extreme cases, such as the least costly product, which could be useful for businesses looking to market entry-level options to customers.
STDDEV	Computes the standard deviation of a specified numeric column, which measures the amount of variation or dispersion of a set of values. Useful in quality control, finance, and any field where variability is key to the analysis.
VAR	Calculates the variance of a specified numeric column, similar to STDDEV, but gives the square of the dispersion. This is critical in financial and scientific calculations where understanding variability is essential.

The Fourth Story: Analysis of Social Media Hashtags

Jacob is a data analyst at a social media company who is responsible for analyzing the hashtag usage in user posts over the past year. A table called HashtagUsage (shown in Table 2-18) logs every hashtag used, the number of likes each post received, and the number of comments each post received.

Table 2-18. *The HashtagUsage Table*

Hashtag	Likes	Comments
#adventure	150	30
#adventure	200	45
#foodie	300	60
#travel	250	50
#foodie	180	20
#staycation	190	40

Jacob aims to answer these four questions using this data:

- How many hashtag entries have been recorded so far?

- For a general idea of how much engagement is going on, how many likes and comments are generated per hashtag?

- What is the maximum engagement rate to the popular likes and comments?

- What is the minimum engagement to identify less popular or ineffective hashtags?

The following query can generate a baseline understanding of how hashtags are utilized:

```
SELECT COUNT(*) AS TotalHashtags
FROM HashtagUsage;
```

This result tells Jacob that there are six entries in the HashtagUsage table. See Table 2-19.

Table 2-19. *Total Hashtags*

TotalHashtags
6

Note   When counting total hashtag entries using the COUNT function, it's important to consider the uniqueness of the values in the Hashtag column. If the Hashtag values are not unique, and you are interested in knowing the number of distinct hashtags used rather than the total number of hashtag occurrences, you should use the DISTINCT keyword within the COUNT() function. For example, COUNT(DISTINCT Hashtag) will count only unique hashtag entries, ensuring that each hashtag is counted only once regardless of how many times it appears in the dataset. This approach is crucial when analyzing data for diversity in hashtag usage, as it provides a more accurate reflection of the range of distinct tags used across posts, rather than merely quantifying their total use.

SELECT COUNT(DISTINCT Hashtag) AS DistinctHashtags

FROM HashtagUsage;

As a result of this query, the number of unique hashtags will be returned, regardless of how many times each appears in the table.

The following query calculates the average number of likes and comments per hashtag to determine the general level of engagement with each hashtag.

```
SELECT AVG(Likes) AS AverageLikes, AVG(Comments) AS AverageComments
FROM HashtagUsage;
```

Table 2-20 shows that on average, hashtags got about 211.67 likes and 40.83 comments.

Table 2-20. *Average Likes and Comments*

AverageLikes	AverageComments
211.67	40.83

To see peak engagement and the maximum likes and comments any hashtag has received, you can view the highest points of user interaction with this query.

```
SELECT MAX(Likes) AS MaxLikes, MAX(Comments) AS MaxComments
FROM HashtagUsage;
```

Table 2-21 indicates that the maximum likes and comments any single hashtag received are 300 and 60, respectively.

Table 2-21. *Maximum Likes and Comments*

MaxLikes	MaxComments
300	60

To understand the least interaction some hashtags receive and the minimum engagement, the following query can identify the less popular or ineffective tags:

```
SELECT MIN(Likes) AS MinLikes, MIN(Comments) AS MinComments
FROM HashtagUsage;
```

Table 2-22 shows that the minimum likes and comments recorded are 150 and 20, respectively.

Table 2-22. *Minimum Likes and Comments*

MinLikes	MinComments
150	20

By using the total hashtag usage data and the queries, Jacob discovers likes (AverageLikes) and comments (AverageComments), as well as the peak of user engagement in posts with likes (MaxLikes), comments (MaxComments), likes (MinLikes) and comments (MinComments).

Data analysts and business intelligence professionals rely on these aggregate functions in SQL. They enable the preparation of reports and analytics that communicate the most relevant statistical highlights of large datasets. As these functions are integrated into data storytelling, they help to create a narrative that is both compelling and based on quantitative evidence, thereby enhancing the informational value of the story. These aggregate functions—as well as other aggregate functions introduced in following chapters—will be used to create narratives in the following chapters.

Summary

The purpose of this chapter was to uncover the profound capabilities of SQL in data storytelling by exploring the SELECT statement and its extensions. This chapter explained how to enhance queries with aliases, combine data creatively using CONCAT, and apply mathematical logic directly within SQL. As a result of the introduction of the CASE and DISTINCT selections, you are now equipped with additional tools for efficiently segmenting and analyzing your data.

Key Points

- SELECT: A SQL command used to retrieve rows selected from one or more tables in a database.

- **Aliases**: Alternative names given to a table or a column in an SQL statement to simplify query readability or to resolve naming conflicts.

- **The CONCAT function**: A function in SQL that merges two or more strings into one, allowing for dynamic string construction within queries.

- **SQL mathematical operations**: Operations that perform regular arithmetic calculations like addition, subtraction, multiplication, and division directly on database columns to manipulate numeric data.

- CASE: A conditional expression in SQL that allows for different outputs in a query, based on specified conditions, similar to if-then logic in programming.

- DISTINCT: A keyword in SQL used to return only distinct (different) values within a column, eliminating duplicate entries in query results.

- **Aggregate functions**: Functions in SQL that perform a calculation on a set of values and return a single value, commonly used for statistical analysis over data groups, such as SUM( ), AVG( ), MIN( ), MAX( ), and COUNT( ).

During this SQL journey, I encourage you to experiment with these tools as you develop your data projects.

Key Takeaways

- **Versatility of** SELECT: The SELECT statement is the cornerstone of SQL querying, allowing you to retrieve precisely the data you need from a database.

- **Using aliases**: Aliases simplify queries and improve the readability of the results. They are especially useful in reports and complex queries involving multiple tables or when column names are lengthy or not understandable.

- **The power of** CONCAT: The CONCAT function is a powerful tool for merging columns and text within SQL queries, enabling the creation of new, meaningful strings that can enhance data interpretation and storytelling.

- **Mathematical operations**: SQL allows for direct mathematical manipulations within the SELECT statement, facilitating immediate calculations.

- **Conditional logic with** CASE: The CASE statement enriches SQL queries by introducing conditional logic directly into SELECT statements.

- **Importance of distinct selection**: Using DISTINCT helps ensure that query results contain unique data points, which is crucial for accurate reporting and analysis, avoiding data redundancy and skewing.

- **Aggregate functions**: Functions like COUNT, AVG, MAX, MIN, and others provide statistical insights directly from the database and are crucial for summarizing data and drawing meaningful conclusions for strategic decision-making.

You will learn advanced SQL topics as you progress through the chapters. This will enable you to develop insights that help you make decisions.

Looking Ahead

As you move to the next chapter, "Filtering Facts with WHERE," you will learn about another of SQL's most frequently used commands. This chapter explains how to craft precise WHERE statements to filter data based on specific conditions. It will allow you to refine your datasets and extract even more targeted insights, which will enable you to tell deeper stories with your data.

Test Your Skills

1. A popular video rental store wants to create a report displaying movie titles, release years, and rental categories. The classification of some movies has not yet been completed. Write a query that retrieves the Title and ReleaseYear and adds a column named Category. Using CASE, it is possible to assign categories according to release year. Movies released before 1990 are Classic; movies released from 1991 to 1999 are the 90s; and movies released after 2000 are labeled Twenty-first century.

2. A fitness instructor wants to analyze workout data using a table that stores UserID, DistanceWalked, and TimeSpentWalking. For each user, they want to calculate the average walking speed, distance divided by time. Write a query that retrieves the UserID,

DistanceWalked, and TimeSpentWalking, and adds a column named AverageSpeed, calculated by dividing DistanceWalked by TimeSpentWalking.

3. An online clothing store manager uses a database to track the OrderID, CustomerID, and OrderValue. The manager wants to know the total number of orders, the average order amount, and the order with the highest total. Write a query to satisfy this request.

CHAPTER 3

Filtering Facts with WHERE

The WHERE clause is a crucial feature of SQL. It specifies the conditions under which rows should be selected, updated, or deleted. As a result of this statement, the records returned from a query are filtered based on whether certain conditions are true. This allows you to refine your searches to a subset of data within a larger dataset, so that operations are only performed on data entries that meet specific criteria.

Introduction to WHERE

As mentioned earlier, a SQL statement is the first step in building a query. The process of asking SQL to perform a specific task begins with the creation of a statement. As mentioned earlier, the four basic statements in SQL are SELECT, UPDATE, DELETE, and INSERT. A *clause* is a component of a SQL statement that specifies conditions or modifies the statement's action. Data can be filtered, sorted, or grouped using clauses. Clauses specify what data is retrieved or manipulated and how it should be processed. One of the commonly used clauses, WHERE, is discussed in this chapter. The WHERE clause acts like a filter, allowing you to select only specific records from a table that meet certain conditions. This is incredibly useful for narrowing down large datasets to find the information you need.

The Importance of WHERE in Storytelling with Data

In the process of effective data storytelling, the WHERE clause plays a fundamental part in crafting narratives that are based on specific parts of the data. The clause enables storytellers to zoom in on data that reveals patterns and create targeted narratives that are more engaging and informative.

51

© Hamed Tabrizchi 2025
H. Tabrizchi, *Narrative SQL*, https://doi.org/10.1007/979-8-8688-1560-7_3

The Anatomy of a WHERE clause

The WHERE clause typically follows the FROM clause in a SQL statement. It can include multiple conditions combined with logical operators such as AND, OR, and NOT. Each condition in the WHERE clause can also use functions and subqueries to further refine the filtering criteria. This allows for complex and dynamic data retrieval strategies tailored to specific storytelling needs. Basically, it has the following syntax:

```
SELECT column1, column2, ...
FROM table_name
WHERE condition
```

Where column1, column2, ... are columns or fields that you want to retrieve, table_name is the name of the table from which the data is retrieved, and condition is the criteria for a row to be included in the result set. This condition can include comparisons like equals (=), not equals (<> or !=), greater than (>), less than (<), and many others. Table 3-1 summarizes the most commonly used comparison operators in the WHERE clause. They are essential for filtering data based on specific criteria by comparing column values and constants or between columns themselves.

Table 3-1. *SQL Comparison Operators for the WHERE Clause*

Operator	Description	Example Use
=	Equal to	WHERE age = 30
!=	Not equal to	WHERE age != 30
	Not equal to	WHERE age <> 30
>	Greater than	WHERE age > 30
<	Less than	WHERE age < 30
>=	Greater than or equal to	WHERE age >= 30
<=	Less than or equal to	WHERE age <= 30
BETWEEN	Between an inclusive range	WHERE age BETWEEN 25 AND 35
LIKE	Search for a pattern	WHERE name LIKE 'J%'
IN	Match any of a list of values	WHERE age IN (20, 30, 40)
IS NULL	Matches if the column is NULL	WHERE name IS NULL

The First Story: The Online Shop

Alex manages an online shop that has experienced rapid growth over the past year. As the business expands, Alex finds it increasingly important to make data-driven decisions to enhance customer satisfaction. He needs to analyze various aspects of the shop's performance, from understanding customer behavior to monitoring order statuses. The online shop has a database that contains a table called Orders. Table 3-2 illustrates the Orders table.

Table 3-2. *The Orders Table*

OrderID	CustomerID	OrderDate	TotalAmount	Status
1	10	2023-05-01	120.00	Delivered
2	20	2023-05-02	75.00	Shipped
3	10	2023-05-03	200.00	Pending
4	30	2023-05-04	150.00	Canceled
5	20	2023-05-05	500.00	Delivered
6	30	2023-05-01	60.00	Pending

To analyze various aspects of the shop's performance, Alex is seeking answers to the following questions:

- Alex wants to find all orders with a total amount greater than $100.

- He wants to select all orders with a status of Shipped.

- He needs to find all orders placed on a specific date, 2023-05-01.

- He needs to find all orders with a total amount greater than $200 and with the Delivered status.

- He wants to select all orders that are either in Pending status or have a total amount greater than $500.

- He wants to find all orders placed on 2023-05-01 with a total amount between $50 and $150.

The following SQL queries will be used to answer each of these questions.

First, the following query retrieves all orders with a total amount greater than $100. A query such as this may be useful in identifying high-value transactions, which might indicate significant purchases or significant customers:

```
SELECT *
FROM Orders
WHERE TotalAmount > 100;
```

Table 3-3 shows the resulting table, which provides all orders with a total amount greater than $100.

Table 3-3. *All Orders with a Total Amount Greater Than $100*

OrderID	CustomerID	OrderDate	TotalAmount	Status
1	10	2023-05-01	120.00	Delivered
3	10	2023-05-03	200.00	Pending
4	30	2023-05-04	150.00	Canceled
5	20	2023-05-05	500.00	Delivered

The following query can be used to filter orders that are currently in the Shipped status. By using this query, it is possible to track orders that are on their way to customers but have not yet been delivered:

```
SELECT *
FROM Orders
WHERE Status = 'Shipped';
```

Table 3-4 illustrates orders that have been loaded.

Table 3-4. *Orders That Have Been Loaded*

OrderID	CustomerID	OrderDate	TotalAmount	Status
2	20	2023-05-02	75.00	Shipped

The following query retrieves all orders placed on May 1, 2023. For the purpose of analyzing sales activity on a particular day, it is useful.

```
SELECT *
FROM Orders
WHERE OrderDate = '2023-05-01';
```

Table 3-5 shows all the orders placed on May 1, 2023.

Table 3-5. *Orders Placed on May 1, 2023*

OrderID	CustomerID	OrderDate	TotalAmount	Status
1	10	2023-05-01	120.00	Delivered
6	30	2023-05-01	60.00	Pending

Dates play a crucial role in data analysis with SQL, providing insight into trends, patterns, and behaviors over time. The results presented in Table 3-5 provide Alex with an opportunity to analyze date-based data for his next analytical task.

Note   Dates are crucial in SQL for filtering, sorting, and aggregating time-based data. An accurate and meaningful query result depends on the proper handling of dates. It should be noted that SQL provides various functions to manipulate dates, including `NOW()`, `CURDATE()`, `DATEADD()`, `DATEDIFF()`, and `DATE_FORMAT()`. These functions help perform operations like getting the current date, adding or subtracting dates, calculating differences, and formatting dates. Also, SQL mostly stores dates in a standard format, such as `YYYY-MM-DD`, to avoid inconsistencies and ensure compatibility across different systems. However, it is possible to use date functions to extract parts of dates for comparison. For example, you can use `YEAR(OrderDate) = 2023` to filter records by year.

There are a number of common mistakes when dealing with dates in SQL. It is possible to have format inconsistencies when storing dates as strings. Thus, it is highly recommended to use proper date data types like `DATE`, `DATETIME`, and `TIMESTAMP`. Additionally, you must be careful when dealing with timezones and using either a `TIMESTAMP` or a `DATETIME`. In order to avoid discrepancies, it is critical that the database and application are aligned on the same time zone.

Another common error when dealing with dates in SQL is comparing dates. It is possible to obtain unexpected results if the time component is ignored. For instance, `WHERE OrderDate = '2023-05-01'` might miss records with timestamps. It is recommended to use `DATE(OrderDate) = '2023-05-01'` to ignore the time part.

The following query considers multiple conditions to retrieve all orders with a total amount greater than $200 that were received by customers. This query identifies orders with a total amount greater than $200 placed by customers who received their order. This allows Alex to segment his high-value customers.

```
SELECT *
FROM Orders
WHERE TotalAmount > 200 AND Status = 'Delivered';
```

Table 3-6 shows all received orders with a total amount greater than $200.

Table 3-6. *Received Orders with a Total Amount Greater Than $200*

OrderID	CustomerID	OrderDate	TotalAmount	Status
5	20	2023-05-05	500.00	Delivered

The following query finds all orders that are either in `Pending` status or have a total amount greater than $500, thus highlighting ongoing or exceptionally large transactions.

```
SELECT *
FROM Orders
WHERE Status = 'Pending' OR TotalAmount > 500;
```

Table 3-7 indicates all orders that are either in `Pending` status or have a total amount greater than $500.

Table 3-7. *Orders That Are in Pending Status or Have a Total Amount Greater Than $500*

OrderID	CustomerID	OrderDate	TotalAmount	Status
3	10	2023-05-03	200.00	Pending
6	30	2023-05-01	60.00	Pending

The following query filters orders made on May 1, 2023, where the total amount is between $50 and $150. This query is useful in analyzing moderate transactions on a particular date:

```
SELECT *
FROM Orders
WHERE OrderDate = '2023-05-01' AND TotalAmount BETWEEN 50 AND 150;
```

In SQL, the BETWEEN operator is used to filter the result set based on a specified range of values. Depending on the context, this range may include numeric, date, or even text values. As the BETWEEN operator is inclusive, it includes both the start and end values. For instance, here, TotalAmount is the column we are filtering, and 50 and 150 define the range. The result will include rows where TotalAmount has values between 50 and 150, inclusive.

Note   Using BETWEEN makes queries easier to read and understand than when using multiple conditions (e.g., >= and <=). BETWEEN includes boundary values. To create an exclusive range, you must use < and >. Additionally, it is necessary to ensure that the data types of the column and the values are the same. It is possible to encounter unexpected results or errors when data types are out of alignment.

It is important to point out some common mistakes associated with the use of BETWEEN. When dealing with ranges in a large dataset, forgetting that BETWEEN includes both endpoints can lead to off-by-one errors. Also, using incorrect date formats can lead to wrong results or errors. Always use the proper date format supported by your SQL database system.

Table 3-8 shows moderate transactions (between $50 and $150) as of May 1, 2023.

Table 3-8. *Transactions Between $50 and $150 on May 1, 2023*

OrderID	CustomerID	OrderDate	TotalAmount	Status
1	10	2023-05-01	120.00	Delivered
6	30	2023-05-01	60.00	Pending

In the process of executing the queries, Alex gained valuable insight into the operation and behavior of the online shop. He identified high-value transactions by filtering orders, highlighting significant purchases that contributed substantially to revenue. Through the examination of orders placed on specific dates, he was able to identify potential promotional success dates, such as May 1, 2023, which was a significant shopping day.

The analysis of order statuses revealed the efficiency of the shipping process and areas needing improvement, with insights into pending and delivered orders. Combining multiple conditions in queries allowed Alex to determine specific customer segments, such as high-value customers, and understand their purchasing patterns. These insights empowered Alex to make data-driven decisions, optimize inventory management, enhance customer satisfaction, and tailor marketing strategies to boost overall sales and operational efficiency.

Advanced Filtering

It is often difficult to interpret data, regardless of how vast it is. Fortunately, SQL offers powerful tools that enable you to unlock its mysteries. In this section, you learn about advanced filtering techniques.

Using WHERE with Dates

Filtering records based on date conditions is a fundamental skill in SQL, especially crucial for handling time-series data. Dates are often used in reporting, trend analysis, and forecasting. The WHERE clause enables precise filtering by specific dates or ranges, allowing meaningful insights from temporal data. Table 3-9 outlines the key aspects of working with dates in SQL, including the most common functions and operations.

Table 3-9. *SQL Date Handling*

Operation	Description	Example Use
CURRENT_DATE	Returns the current date.	SELECT CURRENT_DATE;
CURRENT_TIMESTAMP	Returns the current date and time.	SELECT CURRENT_TIMESTAMP;
DATE()	Extracts the date part of a date/time expression.	SELECT DATE(OrderDate) FROM Orders;
YEAR()	Extracts the year part of a date.	SELECT YEAR(OrderDate) FROM Orders;
MONTH()	Extracts the month part of a date.	SELECT MONTH(OrderDate) FROM Orders;
DAY()	Extracts the day part of a date.	SELECT DAY(OrderDate) FROM Orders;
DATEDIFF()	Calculates the difference between two dates.	SELECT DATEDIFF('2023-05-10', '2023-05-01');
DATE_ADD()	Adds a specified time interval to a date.	SELECT DATE_ADD('2023-05-01', INTERVAL 7 DAY);
DATE_SUB()	Subtracts a specified time interval from a date.	SELECT DATE_SUB('2023-05-01', INTERVAL 7 DAY);
STR_TO_DATE()	Converts a string to a date.	SELECT STR_TO_DATE('01,05,2023', '%d,%m,%Y');
DATE_FORMAT()	Formats a date based on a specified format.	SELECT DATE_FORMAT(OrderDate, '%W %M %Y') FROM Orders;
BETWEEN	Checks if a date falls within a specified range.	SELECT * FROM Orders WHERE OrderDate BETWEEN '2023-01-01' AND '2023-12-31';
TIMESTAMPDIFF()	Calculates the difference between two dates or date/time values in the specified unit. For instance, seconds, minutes, hours, days, weeks, months, and years.	SELECT TIMESTAMPDIFF(DAY, '2023-01-01', '2023-12-31');

Beyond Exact Matching

Suppose you are searching for a book title but can only recall a fragment of it. You can use the LIKE operator in SQL to match patterns by using wildcards to capture partial matches within text data. In this way, you can uncover relevant information even when your search terms are less than precise, making text searches easier. An overview of pattern matching in SQL is provided in Table 3-10, with a focus on wildcards and the LIKE operator.

Table 3-10. *SQL Pattern Matching Summary*

Pattern Matching Technique	Description	Example	Result
% (percent)	Matches any sequence of characters (including zero characters).	WHERE Name LIKE 'J%'	Names starting with 'J', like 'John', 'Jane', and 'Jack'.
_ (underscore)	Matches exactly one character.	WHERE Name LIKE 'J_n'	Names like 'Jon', 'Jan', but not 'John'.
[charlist]	Matches any single character within the specified range or set.	WHERE Name LIKE 'J[aeiou]n'	Names like 'Jan', 'Jen','Jin', 'Jon',and 'Jun'.
[^charlist] or [!charlist]	Matches any single character not within the specified range or set.	WHERE Name LIKE 'J[^aeiou]n' or WHERE Name LIKE 'J[!aeiou]n'	Names like 'Jyn', 'Jhn',and 'Jzn'.
%...%	Matches any sequence of characters, both before and after the specified pattern.	WHERE Name LIKE '%son%'	Names containing 'son', e.g., 'Johnson', 'Jackson'.
%_	Combines % and _ to match any sequence followed by exactly one character.	WHERE Name LIKE '%_n'	Names ending with 'n' and having at least one preceding character.

Subquery Filtering

Subqueries allow you to create highly specific filtering criteria. A subquery can handle complex filtering by dynamically generating the filter condition based on the subquery's results. Subqueries empower you to explore data from multiple angles. In order to keep your main query clean and easy to understand, you can encapsulate complex filtering logic within a subquery.—in the case of finding products with a higher sales volume than the previous year, for example. The average can be calculated by a subquery and then filtered using that value.

```
SELECT product_id, sales_volume, year
FROM sales_data
WHERE sales_volume > (
  SELECT AVG(sales_volume)
  FROM sales_data
  WHERE year = 2023
) AND year = 2024
```

In essence, this query first calculates the average sales volume for the previous year and then filters the current year's sales data to show only products that have exceeded that average. By assuming that the `sales_data` table has columns like `product_id` (a unique identifier for each product), `sales_volume` (the number of units sold for each product in a specific year), and `year` (the year of the sales data), this nested query can find all the products sold this year with a higher sales volume than last year's average.

The Second Story: A Football Academy

A new batch of young football players joined the club. However, with this number of players on hand, how can the coaches identify the most promising players? Maria, the club's data analyst, wants to analyze the data obtained by football club talent scouts. This includes `Player_id`, `name`, `age`, `position` (the playing position, such as `Defender`, `Midfielder`, `Striker`, or `Goalkeeper`), `goals_or_saves` (the number of goals scored in practice matches or saves made by the goalkeeper), `assists`, and `games_played` (the total number of practice matches played). Table 3-11 shows the data obtained by football club talent scouts.

Table 3-11. *Players Table: Young Football Players Data*

player_id	Name	Age	Position	goals_scored	Assists	games_played
1	Alex Jones	16	Midfielder	8	3	12
2	Ben Miller	18	Defender	1	2	10
3	Charlie Brown	17	Forward	12	1	15
4	David Lee	15	Midfielder	5	4	8
5	Emily Garcia	16	Defender	0	1	7
6	Faye Williams	17	Forward	7	2	11
7	George Smith	18	Midfielder	4	5	9
8	Hannah Davis	15	Defender	2	0	5
9	Isabella Moore	16	Forward	9	3	14
10	Jack Robinson	18	Defender	3	1	12
11	Kevin Thomas	17	Midfielder	6	4	10
12	Lily Johnson	15	Forward	4	2	8

The questions Maria is looking to answer are as follows:

1. How many U17 players (players under 17 years of age) have scored more than five goals?

2. Are there any players who consistently play (games_played >= 10) and have a high goal scoring average (goals_scored / games_played > 0.5)?

3. Which players have scored more goals than the average number of goals scored by all players?

4. Which players are younger than 17 and have more assists than the average assists of players aged 18?

To answer the question "How many U17 players have scored more than five goals?", Maria needs to filter the Players table to include only those players who are under 17 and have scored more than five goals. Afterward, she can count how many players there are.

```
SELECT COUNT(*)
FROM Players
WHERE age < 17 AND goals_scored > 5;
```

WHERE age < 17 is a condition that filters players who are under 17 years of age, and goals_scored > 5 is a condition that further filters players who have scored more than five goals. SELECT COUNT(*) counts the number of players who meet these conditions.

Running the query on the given table will yield the number of U17 players who have scored more than five goals. Based on the provided data, the players who meet these criteria are Alex Jones (16 years old, 8 goals scored) and Isabella Moore (16 years old, 9 goals scored). Therefore, the number of U17 players who have scored more than five goals is two.

To find players who have played at least ten games and have a goal scoring average greater than 0.5, Maria would use this query:

```
SELECT name, goals_scored, games_played, (goals_scored / games_played) AS
goal_scoring_average
FROM Players
WHERE games_played >= 10 AND (goals_scored / games_played) > 0.5;
```

Based on the games_played >= 10 and (goals_scored / games_played) > 0.5 conditions, the players who meet these criteria are shown in Table 3-12. This table provides the names, goals scored, games played, and calculated goal-scoring averages for the players who satisfy the query's conditions.

Table 3-12. *Players Who Have Played at Least Ten Games and Have a Goal Scoring Average Greater Than 0.5*

Name	goals_scored	games_played	goal_scoring_average
Alex Jones	8	12	0.66
Charlie Brown	12	15	0.80
Faye Williams	7	11	0.63
Isabella Moore	9	14	0.64
Kevin Thomas	6	10	0.60

During this time, a player named Jack Robinson asked Maria why he was not included on the list. Maria explains the reason for not choosing him. This result is obtained by applying the conditions to Jack Robinson's data. Condition 1 is games_played >= 10. Jack Robinson played 12 games, so this condition is satisfied. Condition 2 is (goals_scored / games_played) > 0.5. Jack Robinson scored three goals in 12 games. His goal-scoring average is 3/12=0.25. Since the goal-scoring average of 0.25 is not greater than 0.5, Jack Robinson does not meet the second condition. Therefore, he is not included in the query resulting table. This illustrates the importance of both conditions being met for a player to be included in the query results.

The following nested queries can be used to identify players who have scored more goals than the average number of goals scored by each player:

```
SELECT name, goals_scored
FROM Players
WHERE goals_scored > (SELECT AVG(goals_scored) FROM Players);
```

The main query selects the players' names and goals scored. The subquery, SELECT AVG(goals_scored) FROM Players, calculates the average number of goals scored by all players in the Players table. The WHERE clause in the main query checks if the goals_scored by each player is greater than the average goals scored calculated by the subquery. See Table 3-13.

Table 3-13. *Players Who Have Scored More Goals Than the Average Number of Goals Scored by All Players in the Academy*

Name	goals_scored
Alex Jones	8
Charlie Brown	12
Faye Williams	7
Isabella Moore	9
Kevin Thomas	6

As shown in Table 3-13, the result of this query shows players who have scored more goals than the average number of goals scored by all players in the academy.

Note   By using nested SQL queries, it is possible to execute subqueries within the WHERE clause. Data can be filtered based on complex conditions derived from other parts of the same table in this manner. A powerful and insightful query can be created by using subqueries to calculate dynamic values (such as averages).

The following query can be used to identify players who are younger than 17 and have more assists than average compared to players aged 18 and older:

```
SELECT name, age, assists
FROM Players
WHERE age < 17 AND assists > (SELECT AVG(assists) FROM Players WHERE
age = 18);
```

In the main query, names, age, and assists are selected. The subquery, SELECT AVG(assists) FROM Players WHERE age = 18, determines the average number of assists provided by players aged 18. In the main query, the WHERE clause ensures that only players who younger than 17 and who have more assists than the average assists for 18-year-olds are selected. Table 3-14 shows the result of this query, identifying the players.

Table 3-14. *Players Who Are Younger Than 17 and Have More Assists Than the Average Number of Assists Recorded by Players Aged Exactly 18*

Name	Age	Assists
Alex Jones	16	3
David Lee	15	4
Isabella Moore	16	3

Common Mistakes When Using WHERE in SQL and How to Avoid Them

This section introduces common errors associated with the use of the WHERE clause. Querying using WHERE clauses requires careful attention to data type, null values, and case sensitivities.

Data Type Issues

In SQL, filtering data based on different data types is a common task. The WHERE clause can result in errors or unexpected results if data types are handled incorrectly.

For instance, imagine a scenario where a company wants to filter employees based on their start date. The employee data is stored in a table named Employees, with the StartDate column as a string instead of a date type. Here is the wrong query that caused the mistake.

```
SELECT * FROM Employees WHERE StartDate = '2023-05-01'
```

Comparing string representations of dates can be problematic, especially if the format of the dates varies. For instance, '2023-5-1' and '2023-05-01' will be treated as different, leading to missed results.

The solution is to convert the StartDate column to a proper date type before filtering:

```
SELECT * FROM Employees WHERE CAST(StartDate AS DATE) = '2023-05-01'
```

Casting StartDate to a date type ensures that the comparison is accurate and consistent, preventing potential mismatches from occurring.

Note   The CAST() function in SQL is used to convert a value from one data type to another. This is particularly useful when dealing with data that may be stored in one format but needs to be processed or compared in another. The CAST() function helps ensure that data types match appropriately in operations, preventing errors and ensuring accurate results. The basic syntax of the CAST() function is as follows:

CAST(*expression* AS *target_data_type*)

The expression is the value or column you want to convert, and the target_ data_type is the data type to which you want to convert the expression.

There is also a problem with numerical comparisons. Assume that the Salary column is stored as a string.

```
SELECT * FROM Employees WHERE Salary > '50000'
```

As string comparison does not work the same way as numeric comparison, this query may not return the correct results. A solution is to convert the Salary column to a numeric type before filtering it.

```
SELECT * FROM Employees WHERE CAST(Salary AS DECIMAL) > 50000
```

Data type conversion ensures that numeric comparisons are performed correctly, producing the expected results.

Note   DECIMAL is a data type used in SQL to store fixed-point numbers with a guaranteed level of precision and scale. Unlike floating-point numbers (e.g., FLOAT), which can lose accuracy due to internal representation, DECIMAL offers an exact representation of decimal values. The decimal data type is ideal for storing financial data, measurements, or any scenario where precise decimal values are crucial. DECIMAL is often defined with two parameters in parentheses: precision (p) and scale (s).

- Precision (p) represents the total number of digits the number can hold, including digits before and after the decimal point.

- Scale (s) represents the number of digits allowed to the right of the decimal point.

As an example, DECIMAL(5,2) specifies a precision of five digits, allowing for numbers such as 123.45 and -98.76, which have two digits following the decimal point.

Logical Mistakes in Conditions

In the WHERE clause, logical errors can significantly affect the performance of the query. Common logical mistakes include incorrect use of AND and OR, and misunderstanding operator precedence. The AND and OR operators combine conditions to simplify data filtering.

Consider a table named Members that contains the following columns: MemberID, FirstName, LastName, MembershipType (e.g., Standard, Premium), MonthlyFee, JoinDate, and Active (indicating whether the membership is currently active).

The manager of a gym wants to identify members who either have a premium membership or pay more than $40 monthly, but they only want to see members who currently have an active membership.

```
SELECT *
FROM Members
WHERE MembershipType = 'Premium' OR MonthlyFee > 40 AND Active = TRUE
```

There is a logical error here due to a misunderstanding of operator precedence. SQL evaluates AND before OR. Thus, the query is interpreted as follows:

```
SELECT *
FROM Members
WHERE (MembershipType = 'Premium') OR (MonthlyFee > 40 AND Active = TRUE)
```

This returns all premium members, regardless of whether they are active, as well as any members with a monthly fee over $40 who are active. Due to the inclusion of inactive premium members, this result is not as expected.

```
SELECT *
FROM Members
WHERE (MembershipType = 'Premium' OR MonthlyFee > 40) AND Active = TRUE
```

The addition of parentheses ensures that the intended logic is followed. The query now filters members who are either premium members or pay more than $40 per month, and it also checks that they are active. As a result of this approach, accurate and expected results are obtained, while inactive members are filtered out.

Note   There are two types of errors that can occur when writing SQL queries: syntax errors and semantic errors.

- **Syntax errors**: These are errors in the SQL statement form or structure. They are easy to detect because the SQL engine throws an error message indicating the problem. For example, missing a comma, incorrect keyword use, or unmatched parentheses.

- **Semantic errors**: These are logical errors where the query is syntactically correct but does not produce the intended result. Semantic errors are difficult to detect because the SQL engine executes the query without any error messages, but the data returned is incorrect or misleading.

In particular, semantic errors can be particularly problematic since they can go unnoticed, resulting in incorrect data analysis and incorrect decision-making. Syntax errors result in the query not being run, whereas semantic errors result in valid SQL queries producing incorrect or unexpected results.

NULL Handling

Handling NULL values in a WHERE statement is another common source of errors. NULL represents unknown or missing data, and comparisons involving NULL can yield unexpected results.

In SQL, NULL is not equal to anything, not even another NULL. Therefore, special handling is required to correctly filter NULL values.

Consider an Employees table that contains these columns: EmployeeID, FirstName, LastName, Department, and PhoneNumber (which may contain NULL values). An analysis of the data is intended to filter Employees without phone numbers, which refers to those employees who have not indicated their phone number on their application. However, this query contains a logical error.

```
SELECT *
FROM Employees
WHERE PhoneNumber = NULL
```

This query returns no results because NULL is not equal to anything, including another NULL. The comparison PhoneNumber = NULL is always false. The following query contains the correct logic.

```
SELECT *
FROM Employees
WHERE PhoneNumber IS NULL
```

Using IS NULL correctly identifies rows where the PhoneNumber column is NULL. This query returns all employees who have not specified their phone number.

Note   The NULL value in SQL represents data that is missing, unknown, or inapplicable. It is a blank field indicating that a value does not exist in a particular field. NULL is not equivalent to zero or an empty string. It is a distinct marker that indicates the absence of value. NULL cannot be compared directly using standard comparison operators (e.g., = and !=). Instead, specialized operators like IS NULL and IS NOT NULL are used. Any arithmetic operation or concatenation involving NULL results in NULL. SQL provides specific syntax for checking for NULL values:

Checking for NULL:

```
SELECT * FROM Employees WHERE PhoneNumber IS NULL
```

Checking for Non-NULL:

```
SELECT * FROM Employees WHERE PhoneNumber IS NOT NULL
```

Case Sensitivity

The assumption that case insensitivity applies to string comparisons is another common mistake. It is possible for SQL to behave differently depending on the database in terms of case sensitivity. PostgreSQL is case-sensitive by default. Identifiers, such as table names, column names, and data values, are treated according to their case. For instance, SELECT * FROM Users will not return results for a table named users.

Assume you need to find a user with the last name smith. The following query would miss entries like 'Smith' or 'SMITH', depending on the database's tagging settings.

```
SELECT * FROM Employees WHERE LastName = 'smith'
```

The solution is to use functions or adjust collation settings to ensure case-insensitive comparisons:

```
SELECT * FROM Employees WHERE LOWER(LastName) = 'smith'
```

By converting LastName to lowercase, the query correctly identifies all variations of 'smith'.

Note   Both UPPER and LOWER follow similar syntax. It is important to keep in mind that the order in which your results are sorted might be affected by the use of UPPER or LOWER. Due to the fact that uppercase letters are usually sorted before lowercase letters, the sorted order may differ from the original case-sensitive order.

Summary

As part of an effort to improve data querying skills, this chapter explored the use of the WHERE clause. By understanding the WHERE clause and its related functions, you can write more efficient and accurate SQL queries.

Key Points

- WHERE: Filters records based on specified conditions.

- **Dates in SQL**: Handles date and time data types, allowing accurate date-based queries.

- BETWEEN: Selects values within a given range, including endpoints.

- **SQL pattern matching**: Uses LIKE to search for patterns within text fields.

- CAST: Converts a value from one data type to another to ensure proper comparison and manipulation.

- **Common Mistakes When Using WHERE in SQL**:

  - **Data types**: Ensure compatibility between data types to avoid errors.

  - **NULL values**: Properly handle NULL values as they represent unknown or missing data, which can affect query results.

- **Logical mistakes**: Avoid errors in combining conditions with AND and OR by understanding operator precedence and using parentheses.

- **Case sensitivity**: SQL may be case-sensitive, which may affect text-based conditions and comparisons.

Key Takeaways

- WHERE **clause**: Essential for filtering records based on specified conditions and enabling precise data retrieval.

- **Dates in SQL**: Proper handling of date and time data types is crucial for accurate date-based queries and comparisons.

- BETWEEN **operator**: Useful for selecting values within a specific range, including the endpoints, simplifying range queries.

- **SQL pattern matching**: The LIKE operator allows for searching text fields using patterns, making it easier to find records that match specific criteria.

- CAST **function**: Converts values from one data type to another, ensuring proper comparison and manipulation in SQL queries.

Looking Ahead

The next chapter, "Complex Characters with JOINs," explores the various types of JOIN operations, which are fundamental for combining data from multiple tables. Mastery of this operation will enable you to create complex queries that provide deeper insights and more comprehensive analyses of your datasets.

Test Your Skills

1. A library manager wants to create a report displaying the titles and publication years of books published between 2000 and 2010, inclusive. Additionally, she would like to filter out any titles that contain the word *Guide*. Write a query that retrieves the title and publicationyear for these books.

2. An online marketplace manager wants to analyze the prices of items listed in the Electronics category that are priced between $50 and $500. The prices are stored as strings. Write a query that retrieves the `ItemID`, `ItemName`, and `Price` (cast to a decimal) for items in this category and price range.

3. The owner of a fitness app wants to generate a report on users who have walked more than five kilometers in any single session. The `DistanceWalked` is stored in meters. Retrieve `UserID`, `SessionID`, and `DistanceWalked` (converted to kilometers) for these sessions.

4. A music streaming service wants to identify all the songs that were released prior to the year 2000 and contain the word *Love* in the title. Write a query that retrieves the `SongID`, `SongTitle`, and `ReleaseYear` for these songs.

CHAPTER 4

Complex Characters with JOINs

JOINs in SQL are operations that allow you to combine the rows of two or more tables based on a related column between them. In most real-world applications, analyzing data comprehensively requires more than just considering a single table. For a comprehensive analysis, it is very important and necessary to consider different dimensions, thus, using the JOINs operation allows you to reach a comprehensive analysis. In a relational database, data is often normalized and spread across different tables to reduce redundancy and ensure data integrity. JOINs enable you to bring this scattered data together and provide a unified view that can be used for better analysis and reporting.

Introduction to JOINs

JOINs in SQL are operations that are essential for querying relational databases because they enable you to extract meaningful information from multiple tables by linking their data.

Importance of JOINs in Storytelling with Data

JOINs are critical to data storytelling because they allow analysts to create a narrative by combining related pieces of information scattered across multiple tables. For example, to analyze customer orders, you might need data from three tables: Customers, Orders, and Products. JOINs enables you to join these tables together and extract a coherent dataset

© Hamed Tabrizchi 2025
H. Tabrizchi, *Narrative SQL*, https://doi.org/10.1007/979-8-8688-1560-7_4

that provides insights into customer buying behaviors, product popularity, and sales performance. By using the JOINs operator in your queries, you can achieve the following advantages.

- Combine data from multiple sources

- Perform in-depth analysis with access to all relevant data points

- Create complex queries that can answer more complex questions

The Anatomy of a JOIN Clause

A JOIN clause can be described as a bridge between two or more tables. As mentioned, a relational database will contain several tables, and the possibility of merging the data of these tables based on a specific criterion allows you to retrieve and analyze the information that is in separate tables. This is important because real-world databases often store data in separate tables to improve organization and efficiency. A JOIN clause fundamentally contains the following parts:

1. **JOIN types**: This specifies the type of connection performed. Common types are INNER JOIN, LEFT OUTER JOIN, RIGHT OUTER JOIN, FULL OUTER JOIN, and CROSS JOIN. Each type determines how matching and non-matching rows are handled in the result set.

2. **FROM clause**: This clause specifies the tables being attached. The order of the tables in the FROM clause can sometimes affect query performance, especially with certain JOIN types.

Note   Most modern database systems use a query optimizer that analyzes the JOIN clause and chooses the most efficient way to perform the JOINs. Today's modern database optimizers are intelligent. They consider various factors such as table size, indexes, and JOIN conditions when choosing a JOIN strategy. In many cases, the optimizer can perform JOINs efficiently regardless of the order you specify in the FROM clause. If you suspect that a JOIN clause is causing performance problems, it is recommended that you use the tools provided by your database system to examine the actual execution plan. This reveals the JOIN strategy chosen by the optimizer and highlights any bottlenecks.

---

3. **ON clause**: This clause specifies the condition that determines which rows of each table match. The ON statement usually compares the columns of both tables using comparison operators such as =, <, or >.

The basic syntax of an INNER JOIN, which is the most common type of JOIN, is shown here:

```
SELECT Customers.Name, Orders.OrderID, Orders.OrderDate
FROM Customers
INNER JOIN Orders ON Customers.CustomerID = Orders.CustomerID;
```

This query is an INNER JOIN that joins the Customers and Orders tables. The ON clause specifies that rows of Customers match rows of Orders only when Customer ID values are equal. This query returns a result set containing columns from both tables, but only for the rows in which the same customer ordered (i.e., matching CustomerID values).

It should be noted that in SQL, the dot (.) is used to refer to a specific column in a specific table. When there are columns with the same name in different tables, a dot is used to determine which table a column belongs to, in order to avoid confusion about column names.

Types of JOINs

Despite the specific needs of each query, there are different types of SQL JOINs. Each type of SQL JOIN provides different methods of combining data from multiple tables. The purpose of each type of JOIN is explained next.

INNER JOIN: This is a tool for retrieving data when there is an explicit relationship between tables. This ensures that you only get rows with matches in both tables, which is ideal for tasks like finding customers who have placed an order.

- **Purpose**: Retrieves only the rows that have matching values in both tables.

- **Use case**: When you want to get only the records that have related data in both tables.

LEFT/RIGHT OUTER JOIN: These JOINs are important for including all rows from a particular table, the one listed first in the FROM clause, even if there is no match in another table. They are typically used to maintain data from one table while retrieving related information from another table.

LEFT OUTER JOIN:

- **Purpose**: Retrieves all rows from the left table and the matched rows from the right table. If there is no match, the result is NULL on the right side.

- **Use case**: When you want to get all records from the left table, with the related data from the right table, if available.

RIGHT OUTER JOIN:

- **Purpose**: Retrieves all rows from the right table and the matched rows from the left table. If there is no match, the result is NULL on the left side.

- **Use case**: When you want to get all records from the right table, with the related data from the left table, if available.

FULL OUTER JOIN: This JOIN combines the left and right outer join functionality. It returns all rows from both tables and fills unmatched columns. This is useful when you need a complete picture of data from both tables, regardless of relationships.

- **Purpose**: Retrieves all rows when there is a match in either the left or right table. Rows without a match in one of the tables will have NULLs.

- **Use case**: When you want to get all records from both tables, matching where possible.

CROSS JOIN: This JOIN creates a Cartesian product, which basically multiplies all the rows in one table by all the rows in the other table. While often used for visualization purposes, it can be computationally expensive for large datasets. Typically used sparingly or when you specifically need all possible combinations of rows.

- **Purpose**: Returns the Cartesian product of the two tables. Each row from the first table is combined with all rows from the second table.

- **Use case**: When you need all possible combinations of rows from the tables. This is rarely used unless for specific purposes like generating all combinations for testing.

SELF JOIN: This JOIN allows you to join a table to itself based on a certain condition. It is powerful for finding relationships in a table.

- **Purpose**: JOINs a table with itself.

- **Use case**: When you need to compare rows within the same table.

Using different types of JOINs provides flexibility in structuring the data. They allow you to retrieve data based on specific match criteria, include all rows from a specific table, or even examine all possible combinations. JOINs are a useful tool for creating efficient analytical queries on real-world relational databases. Table 4-1 briefly explains the reason for using each type of JOIN.

Table 4-1. *The Different Types of JOINs*

JOIN Type	Description	Query Result	Sample Query
INNER JOIN	Returns only rows where there's a match in both tables based on the JOIN condition.	Matching rows from both tables	SELECT * FROM Table1 INNER JOIN table2 ON Table1.key = Table2.key
LEFT OUTER JOIN	Includes all rows from the left table, specified first in the FROM clause, and matching rows from the right table. If there's no match in the right table, NULL values are filled for columns from the right table.	All rows from the left table, matching rows from the right table, or NULLs for non-matching rows.	SELECT * FROM Table1 LEFT JOIN table2 ON Table1.key = Table2.key
RIGHT OUTER JOIN	Similar to LEFT JOIN, but includes all rows from the right table, specified first in the FROM clause, and matching rows from the left table. NULLs are filled for non-matching rows in the left table.	All rows from the right table, matching rows from the left table, or NULLs for non-matching rows.	SELECT * FROM Table1 RIGHT JOIN table2 ON Table1.key = Table2.key

(*continued*)

Table 4-1. (*continued*)

JOIN Type	Description	Query Result	Sample Query
FULL OUTER JOIN	Combines the LEFT and RIGHT OUTER JOINs. Includes all rows from both tables, regardless of whether there's a match in the other table. NULLs are filled for non-matching columns.	All rows from both tables with NULLs for non-matching rows.	SELECT * FROM Table1 FULL OUTER JOIN table2 ON Table1.key = Table2.key
CROSS JOIN	Creates a Cartesian product, resulting in all possible combinations of rows from both tables, regardless of any matching criteria. Can generate a large number of rows.	All possible combinations of rows from both tables.	SELECT * FROM Table1 CROSS JOIN Table2
SELF JOIN	Joins a table to itself based on a specified condition. Often used to find relationships within the same table.	Rows from the same table that meet the JOIN condition.	SELECT copy1. column1 AS first_column, copy2.column2 As second_column FROM Table1 copy1 INNER JOIN Table1 copy2 ON copy1.key = copy2.key
LEFT OUTER JOIN with NULL check	Like a LEFT JOIN, but explicitly checks for NULLs to filter results.	Rows from the left table with no matching rows in the right table.	SELECT * FROM Table1 LEFT JOIN Table2 ON Table1.key = Table2.key WHERE Table2.key IS NULL

(*continued*)

Table 4-1. (*continued*)

JOIN Type	Description	Query Result	Sample Query
RIGHT OUTER JOIN with NULL check	Like a RIGHT JOIN, but explicitly checks for NULLs to filter results.	Rows from the right table with no matching rows in the left table.	SELECT * FROM Table1 RIGHT JOIN Table2 ON Table1.key = Table2.key WHERE Table1.key IS NULL
FULL OUTER JOIN with NULL check	Like a FULL JOIN, but explicitly checks for NULLs to filter results.	Rows with no matching rows in the other table.	SELECT * FROM Table1 FULL OUTER JOIN Table2 ON Table1.key = Table2.key WHERE Table1.key IS NULL OR Table2.key IS NULL

Table 4-1 briefly introduced the different types of SQL JOINs. It should be noted that it is necessary to combine rows from two or more tables based on a related column, known as a *key*. In the rest of this chapter, all types of keys are fully explained.

As stated in the table, in summary, an INNER JOIN returns only matching rows from both tables, while a LEFT JOIN includes all rows from the left table and matching rows from the right, filling the right table columns with NULL. Conversely, RIGHT JOIN includes all rows in the right table with NULL for unmatched left table rows. A FULL OUTER JOIN returns all rows from both tables, using NULL for no match. A CROSS JOIN creates a Cartesian product of all rows. A SELF JOIN joins a table to itself to find related data in the same table. The LEFT, RIGHT, and FULL OUTER JOIN special cases with NULL check the results to show unmatched rows. Understanding these JOIN types and keys helps you optimize database queries and data analysis. To achieve a more intuitive understanding, these JOINs are illustrated in Figure 4-1.

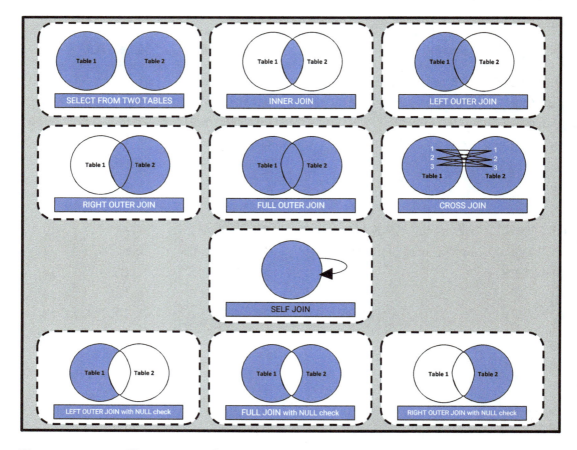

Figure 4-1. *An illustration of JOIN types*

The First Story: A Football Academy

In a football academy, academy data analyst William is under pressure from the club's management staff to find the next young star. Having a database of last season's information, William plans to find the club's next young star by determining the following information:

1. List all the matches along with the names of players who participated.

2. List the players and the total number of goals they scored across all matches.

William's goal is to analyze the data that is stored in the database, including the tables called Player, Matches, and MatchDetails. These tables store each player's statistics and their performance in each match. The Player, Matches, and MatchDetails tables are shown in Tables 4-2, 4-3, and 4-4, respectively.

Table 4-2. *The Player Table*

PlayerID	Name	Age	Position
1	Alex Jones	17	Midfielder
2	Mia Garcia	16	Defender
3	David Lee	18	Forward
4	Sarah Miller	15	Midfielder
5	Chris Brown	17	Defender
6	Emily Sanchez	16	Goalkeeper
7	Ben Johnson	18	Midfielder (Winger)

Table 4-3. *The Matches Table*

MatchID	AgeGroup	MinutesPlayed
101	U18	90
102	U16	45
103	U18	90
104	U16	90
105	U18	90
106	U16	90
107	U18	70

Table 4-4. *The MatchDetails Table*

MatchDetailID	MatchID	PlayerID	GoalsScored
1	101	1	2
2	101	2	2
3	102	2	0
4	103	3	1
5	103	4	2
6	104	4	1
7	105	1	1
8	105	5	0
9	106	6	1
10	107	3	2
11	107	7	1

It should be noted that the Players, Matches, and MatchDetails tables use primary and foreign keys to ensure data integrity and create relationships between tables. The Players table has a primary key called PlayerID, which uniquely identifies each player and ensures that there are no duplicate records for players. The Matches table has a primary key called MatchID, which uniquely identifies each match and ensures that each match record is unique. The MatchDetails table, which links players and matches, has a composite primary key called MatchDetailID, which uniquely identifies each record in the MatchDetails table. Additionally, the MatchDetails table contains two foreign keys: MatchID and PlayerID. The MatchID foreign key references the MatchID in the Matches table, establishing a relationship between the match details and the corresponding matches. The same goes for the Players table. These primary and foreign key constraints guarantee integrity, meaning that each match detail must match valid entries in both the Players and Matches tables.

Note   In a relational database table, a *key* is a column or set of columns that uniquely identifies a row in the table. In other words, this column or columns act like a fingerprint, ensuring that no two rows have the same value for that key. Defining a key in a table is very important to maintaining data integrity and efficiently retrieving specific records. The rest of this chapter explains the keys in the database in more detail.

To find all matches along with the names of players who participated in them, including their positions, an INNER JOIN can be used to match players with their recent performances:

```
SELECT m.MatchID, p.Name, p.Position
FROM Matches m
INNER JOIN MatchDetails md ON m.MatchID = md.MatchID
INNER JOIN Players p ON p.PlayerID = md.PlayerID;
```

This query retrieves a list of matches, along with the names and positions of the players who participated in each match, using two INNER JOIN operations to combine data from the three tables: Matches, MatchDetails, and Players. First, the Matches table (aliased as m) is linked to the MatchDetails table (aliased as md) on the MatchID column, ensuring that each detail is associated with a corresponding match. The resulting dataset is then further linked to the Players table (aliased as p) on the PlayerID column, associating each match detail to the corresponding player. The SELECT statement specifies that the query returns the MatchID from the Matches table and the Name and Position from the Players table. It effectively provides a list of matches, detailing the players who participated in each match and their respective positions, providing a clear view of the players' participation in the various matches.

Note   As mentioned in previous chapters, in SQL, an *alias* is a temporary name given to a table or column for the duration of a query. Aliases are often used to make complex queries more readable and shorten long table names, making SQL code easier to write and understand. To create an alias for a table, you use the AS keyword followed by the alias. For example, in the FROM Players AS p statement, Players is the main table name and p is the alias. It should be noted that the AS keyword is optional, so you can also write FROM Players p. Once an alias is assigned, you can use it to refer to the table in the rest of the query. This is especially useful when dealing with multiple tables or when doing JOINs, as it helps to clearly distinguish between different tables and their columns.

The result of the first query is illustrated in Table 4-5.

Table 4-5. *The First Query Result*

MatchID	Name	Position
101	Alex Jones	Midfielder
101	Mia Garcia	Defender
102	Mia Garcia	Defender
103	David Lee	Forward
103	Sarah Miller	Midfielder
104	Sarah Miller	Midfielder
105	Alex Jones	Midfielder
105	Chris Brown	Defender
106	Emily Sanchez	Goalkeeper
107	David Lee	Forward
107	Ben Johnson	Midfielder (Winger)

To retrieve a list of all players and the total number of goals they scored across all matches, use this query:

```
SELECT p.Name, SUM(md.GoalsScored) AS TotalGoals
FROM Players p
INNER JOIN MatchDetails md ON p.PlayerID = md.PlayerID
GROUP BY p.Name;
```

To return a list of players along with the total number of goals scored by each player, this query performs an INNER JOIN with the MatchDetails table on the PlayerID column to ensure that each player's match details are included. The SUM(md. GoalsScored) function is used to calculate the total goals scored by each player in all their matches. The GROUP BY p.Name statement groups the results by each player's name, so that the sum function (SUM) can be applied to each group. The query returns a set of results containing each player's name and the corresponding total goals they have scored, effectively summarizing each player's scoring performance across all recorded matches.

Note   GROUP BY is a clause used within a SELECT statement in SQL. The GROUP BY clause in SQL arranges identical data into groups. It is often used in conjunction with aggregate functions such as COUNT(), SUM(), AVG(), MAX(), and MIN() to perform calculations on any group of data. The next chapter focuses on this statement in more detail along with several narratives.

The result of the second query is illustrated in Table 4-6.

Table 4-6. *The Second Query Result*

Name	TotalGoals
Alex Jones	3
Mia Garcia	2
David Lee	3
Sarah Miller	3
Chris Brown	0
Emily Sanchez	1
Ben Johnson	1

William was able to get the data he needed by joining the tables together. The questions that William aimed to answer were well-suited for an INNER JOIN, as it only returns rows where there's a match in both tables. This means there wouldn't be any NULL values appearing in the result table. If William wanted to find all players, even players who have not scored a goal yet, and INNER JOIN would not be sufficient, and he would have to use another type of JOIN in his query. He would likely need a LEFT JOIN or a RIGHT JOIN, which would include rows from one table even if there's no corresponding data in the other table. This would inevitably lead to NULL values in certain columns for those unmatched rows. In the next story of this chapter, you learn more about these types of JOINs.

Keys in Relational Databases

Joining tables in SQL opens up powerful data-retrieval capabilities, but careful planning is critical. Choosing the appropriate join type (INNER, LEFT, RIGHT, or FULL) depends on whether you want to match all rows from one or both tables, with or without matches. JOIN conditions, preferably based on primary and foreign keys, ensure accurate results. For this purpose, this section expands your horizons with various keys in relational databases. In relational databases, keys are used to uniquely identify rows within a table and to establish relationships between tables (see Table 4-7).

Table 4-7. *Keys in Relational Databases*

Relational Database Key	Description	Example
Primary Key	A column that uniquely identifies each row in a table.	`PRIMARY KEY (column_name)`
Foreign Key	A column or a set of columns in one table that uniquely identifies rows in another table. Establishes a link between the data in the two tables.	`FOREIGN KEY (column_name) REFERENCES other_table(other_column)`
Unique Key	A column or a set of columns that uniquely identifies each row in a table, but unlike primary keys, a table can have multiple unique keys.	`UNIQUE (column_name`
Composite Key	A primary key composed of multiple columns.	`PRIMARY KEY (column1, column2)`
Candidate Key	A column or a set of columns that can uniquely identify any database record without referring to any other data.	Usually any column or combination of columns that can act as a primary key.
Super Key	A set of one or more columns that can be used to uniquely identify a row in a table. A super key includes the primary key and any additional columns that make it unique.	Usually any primary key or combination of columns that uniquely identifies a row.

To get more familiar with keys, you'll see how to create tables from the previous story by using SQL. In the previous story, there were three tables: Player, Matches, and MatchDetails. These queries can create the Player, Matches, and MatchDetails tables in the database.

```
CREATE TABLE Players (
    PlayerID INT PRIMARY KEY,
    Name VARCHAR(100),
    Age INT,
    Position VARCHAR(50)
);
```

This query creates a table named Players in the database. The CREATE TABLE statement is used to define the structure of the table, specifying the columns and their data types. The PlayerID column is defined as an INT (integer) and is designated as the PRIMARY KEY, which means it will uniquely identify each row in the table. This ensures that no two players can have the same PlayerID. The Name column is defined as a VARCHAR(100), allowing it to store variable-length character strings up to 100 characters in length, suitable for storing player names. The Age column is defined as an INT, which will store the player's age as an integer value. The Position column is defined as a VARCHAR(50), allowing it to store variable-length character strings up to 50 characters in length, which is suitable for storing the player's position on the field.

```
CREATE TABLE Matches (
    MatchID INT PRIMARY KEY,
    AgeGroup VARCHAR(10),
    MinutesPlayed INT
);
```

This query creates a table called Matches in the database. The CREATE TABLE statement defines the structure of the table and its columns and data types. The MatchID column is defined as INT (integer) and set as the PRIMARY KEY to ensure that no two matches have the same MatchID. The AgeGroup column is defined as VARCHAR(10), which allows it to store variable-length character strings of up to ten characters, which is suitable for sorting items by age group. The minutes played column is defined as INT, which stores the total number of minutes played in the match as an integer value.

```
CREATE TABLE MatchDetails (
    MatchDetailID INT PRIMARY KEY,
    MatchID INT,
    PlayerID INT,
    GoalsScored INT,
    FOREIGN KEY (MatchID) REFERENCES Matches(MatchID),
    FOREIGN KEY (PlayerID) REFERENCES Players(PlayerID)
);
```

This query creates a table called `MatchDetails` in the database. The `CREATE TABLE` statement defines the structure of the table, the columns, their data types, and the relationships between this table and the `Matches` and `Players` tables. The `MatchDetailID` column is defined as `INT` (integer) and is set as the `PRIMARY KEY`, which uniquely identifies each row in the table. This ensures that no two match details have the same `MatchDetailID`. The `MatchID` column is defined as `INT`, which stores the match ID. This column is a foreign key that refers to the `MatchID` column in the `Matches` table. This relationship ensures that each entry in `MatchDetails` is a valid match. The `PlayerID` column is defined as `INT`, which stores the player ID. This column is a foreign key that references the `PlayerID` column in the `Players` table. This relationship ensures that each entry in `MatchDetails` is valid for a player. The `GoalsScored` column is defined as `INT`, which stores the number of goals scored by the player in the match.

Note   FOREIGN KEY (column) REFERENCES other_table(other_column) defines a foreign key constraint in the relational database schema.

- FOREIGN KEY (column): This specifies a column or set of columns within the current table that will act as the foreign key. A foreign key references data in another table.

- REFERENCES other_table(other column): This part defines the relationship between the foreign key and another table.

- other_table: This refers to the name of the referenced table that holds the data the foreign key is linked to.

Foreign key constraints ensure referential integrity, meaning that every MatchID in MatchDetails must exist in the Match table, and every PlayerID in MatchDetails must exist in the Player table. This table structure allows accurate tracking of player performance in specific matches, including the number of goals scored by each player in each match. In the rest of the book, you will come across narratives and examples where data is stored by defining composite, candidate, and super keys.

The Second Story: A Technology Company

Piper works for a growing technology company with a dynamic organizational structure consisting of multiple departments, employees, and projects. Piper intends to find the answers of questions about the company using the company data. Piper's company data has been collected and stored in four tables: Employees (see Table 4-8), Departments (see Table 4-9), Projects (see Table 4-10), and Employees_Projects (see Table 4-11).

Table 4-8. *The Employees Table*

EmployeeID	Name	Age	Position	DepartmentID	ManagerID
1	John Smith	45	CEO	NULL	NULL
2	Jane Dylan	38	CTO	1	1
3	Mary Johnson	28	Developer	1	2
4	Mike Brown	35	Developer	1	2
5	Emily Davis	30	HR Manager	2	1
6	Laura Wilson	25	HR Associate	2	5
7	David White	50	CFO	3	1
8	Steve Black	40	Accountant	3	7

Table 4-9. *The Departments Table*

DepartmentID	DepartmentName
1	Engineering
2	Human Resources
3	Finance

Table 4-10. *The Projects Table*

ProjectID	ProjectName	DepartmentID
101	Project Alpha	1
102	Project Beta	1
201	Recruitment Drive	2
301	Financial Audit	3

Table 4-11. *The Employees_Projects Table*

EmployeeID	ProjectID
3	101
4	101
3	102
4	102
6	201
8	301

Piper's purpose for analyzing the data table is to find the answer to the following questions:

- Who are the managers and their direct reports?

- Which departments have employees, and what are their names? List all departments and their employees, including departments without employees.

- Which employees are assigned to which projects, including those
  not assigned to any project? List all employees and their projects,
  including employees not assigned to any project.

- What are the projects and the departments they are assigned to,
  including departments without projects? List all projects and their
  assigned departments, including departments without projects.

- What are all possible combinations of employees and projects?
  Create a Cartesian product of all employees and projects.

- Which employees are not assigned to any project? Find employees
  who are not assigned to any project.

To list the managers and their direct reports, you can use a SELF JOIN on the
Employees table:

```
SELECT e1.Name AS EmployeeName, e2.Name AS ManagerName
FROM Employees e1
LEFT JOIN Employees e2 ON e1.ManagerID = e2.EmployeeID;
```

It should be noted that when you use a LEFT JOIN or RIGHT JOIN (or any type of
JOIN) on the same table, it is called a *self-join*. A *self-join* is simply joining a table with
itself. This is useful for hierarchical or recursive data structures, such as organizational
charts or family trees. Using a LEFT JOIN ensures that all employees are included in
the result set, even those without a manager (the ManagerID is NULL). This way, you can
identify employees who do not report to anyone. See Table 4-12.

Table 4-12. *The First Query Result*

EmployeeName	ManagerName
John Smith	NULL
Jane Dylan	John Smith
Mary Johnson	Jane Dylan
Mike Brown	Jane Dylan
Emily Davis	John Smith
Laura Wilson	Emily Davis
David White	John Smith
Steve Black	David White

To list all departments and their employees, including departments without employees, you use a LEFT JOIN:

```
SELECT d.DepartmentName, e.Name AS EmployeeName
FROM Departments d
LEFT JOIN Employees e ON d.DepartmentID = e.DepartmentID;
```

The query retrieves each department along with the names of employees working in those departments. By using the LEFT JOIN, you can ensure that all departments are included in the result, even if no employees are assigned to some departments. See Table 4-13.

Table 4-13. *The Second Query Result*

DepartmentName	EmployeeName
Engineering	Jane Dylan
Engineering	Mary Johnson
Engineering	Mike Brown
Human Resources	Emily Davis
Human Resources	Laura Wilson
Finance	David White
Finance	Steve Black

To list all employees and their projects, including employees not assigned to any project, you need to use a RIGHT JOIN:

```
SELECT e.Name AS EmployeeName, p.ProjectName
FROM Employees e
RIGHT JOIN Employees_Projects ep ON e.EmployeeID = ep.EmployeeID
RIGHT JOIN Projects p ON ep.ProjectID = p.ProjectID;
```

This query retrieves a list of all projects along with the names of employees assigned to each project. It uses a RIGHT JOIN to ensure that all projects are included in the result set, even if no employees are assigned to them. The first RIGHT JOIN matches employees to their projects through the Employees_Projects table. The second RIGHT JOIN ensures that all projects from the Projects table are included, even if they have no associated employees. The result set will display the EmployeeName and ProjectName, with EmployeeName being NULL for projects without assigned employees. See Table 4-14.

Table 4-14. *The Third Query Result*

EmployeeName	ProjectName
Mary Johnson	Project Alpha
Mike Brown	Project Alpha
Mary Johnson	Project Beta
Mike Brown	Project Beta
Laura Wilson	Recruitment Drive
Steve Black	Financial Audit
NULL	Project Alpha
NULL	Project Beta
NULL	Recruitment Drive
NULL	Financial Audit

In the context of answering the third question, the presence of NULL rows can be problematic, depending on how you interpret the results. Using a RIGHT JOIN with a NULL check can help address the problem of identifying employees not assigned to any project and projects without assigned employees. This approach allows you to filter out unwanted NULL rows and provide a clearer result. Let's go through how this can be done for the relevant questions:

```
SELECT e.Name AS EmployeeName, p.ProjectName
FROM Employees e
RIGHT JOIN Employees_Projects ep ON e.EmployeeID = ep.EmployeeID
RIGHT JOIN Projects p ON ep.ProjectID = p.ProjectID
WHERE e.EmployeeID IS NOT NULL;
```

This query ensures that you only include rows when there is a valid EmployeeID from the Employees table, thus excluding the NULL rows for unassigned projects. See Table 4-15.

Table 4-15. *The Result of Rewriting the Third Query (Unwanted NULL Rows Were Filtered)*

EmployeeName	ProjectName
Mary Johnson	Project Alpha
Mike Brown	Project Alpha
Mary Johnson	Project Beta
Mike Brown	Project Beta
Laura Wilson	Recruitment Drive
Steve Black	Financial Audit

It should be noted that NULL rows usually appear in SQL JOINs when there are unmatched rows between joined tables. Different types of JOINs handle mismatched rows differently, resulting in NULL values for table columns that do not have a corresponding match. In this analysis, you learned about one of the reasons for the emergence of NULL.

Note   The main reason for the occurrence of NULL rows in SQL JOINs is the corresponding mismatch between the connected tables. The different types of JOIN used (LEFT, RIGHT, FULL, or OUTER) lead to different results, so understanding this behavior helps design queries that appropriately handle or avoid NULL values depending on the analytical needs.

- INNER JOIN: No NULLs as only matched rows are included.

- LEFT JOIN: NULLs appear for right table columns when there's no match in the right table.

- RIGHT JOIN: NULLs appear for left table columns when there's no match in the left table.

- FULL OUTER JOIN: NULLs appear for both table columns when there's no match in either table.

- CROSS JOIN: No NULLs inherently, as all combinations are included.

- SELF JOIN: NULLs appear when a row has no match in the same table.

- LEFT JOIN **with** NULL **check**: Filters out rows with NULLs in the JOIN condition, focusing on non-matching rows.

- RIGHT JOIN **with** NULL **check**: Filters out rows with NULLs in the JOIN condition, focusing on non-matching rows.

- FULL OUTER JOIN **with** NULL **check**: Specifically handles NULLs to highlight unmatched rows from both tables.

It should be noted that there are other reasons for the occurrence of NULL rows, including missed matches and data integrity.

- **Missed matches:** When there are no corresponding entries in the linked table, SQL fills the missing values with NULL to indicate the absence of data.

- **Data integrity:** Incomplete data, such as employees without projects or projects without employees, will result in NULLs in the result set when performing certain JOINs.

The last part of this chapter discusses the emergence of NULL in more detail.

To find all projects and their assigned departments, including departments without projects, with the experience you have now, you can write a query that ensures that you only include rows for which there is a valid `ProjectID` in the `Projects` table, so `NULL` rows are removed for departments without projects.

```
SELECT p.ProjectName, d.DepartmentName
FROM Projects p
RIGHT JOIN Departments d ON p.DepartmentID = d.DepartmentID
WHERE p.ProjectID IS NOT NULL;
```

This query ensures that you only include rows where there is a valid `ProjectID` in the `Projects` table, thus excluding the `NULL` rows for departments without projects. See Table 4-16.

Table 4-16. *The Result of the Fourth Query*

ProjectName	DepartmentName
Project Alpha	Engineering
Project Beta	Engineering
Recruitment Drive	Human Resources
Financial Audit	Finance

By applying the `NULL` check in the `WHERE` clause, you can filter out the rows where there are no valid matches, leading to more meaningful and interpretable results.

To create a Cartesian product of all employees and projects, you can use a `CROSS JOIN` to create a Cartesian product, which generates all possible combinations of employees and projects:

```
SELECT e.Name AS EmployeeName, p.ProjectName
FROM Employees e
CROSS JOIN Projects p;
```

This query generates a Cartesian product of the `Employees` and `Projects` tables, pairing every employee with every project. The `CROSS JOIN` operation ensures that all possible combinations of employees and projects are included in the result set (see Table 4-17).

Table 4-17. *The Result of the Fifth Query*

EmployeeName	ProjectName
John Smith	Project Alpha
Jane Dylan	Project Alpha
Mary Johnson	Project Alpha
Mike Brown	Project Alpha
Emily Davis	Project Alpha
Laura Wilson	Project Alpha
David White	Project Alpha
Steve Black	Project Alpha
John Smith	Project Beta
Jane Dylan	Project Beta
Mary Johnson	Project Beta
Mike Brown	Project Beta
Emily Davis	Project Beta
Laura Wilson	Project Beta
David White	Project Beta
Steve Black	Project Beta
John Smith	Recruitment Drive
Jane Dylan	Recruitment Drive
Mary Johnson	Recruitment Drive
Mike Brown	Recruitment Drive
Emily Davis	Recruitment Drive
Laura Wilson	Recruitment Drive
David White	Recruitment Drive
Steve Black	Recruitment Drive
John Smith	Financial Audit

(*continued*)

Table 4-17. (*continued*)

EmployeeName	ProjectName
Jane Dylan	Financial Audit
Mary Johnson	Financial Audit
Mike Brown	Financial Audit
Emily Davis	Financial Audit
Laura Wilson	Financial Audit
David White	Financial Audit
Steve Black	Financial Audit

To find employees who are not assigned to any project, you use a LEFT JOIN and then check for NULLs to find employees who are not assigned to any project:

```
SELECT e.Name AS EmployeeName
FROM Employees e
LEFT JOIN Employees_Projects ep ON e.EmployeeID = ep.EmployeeID
WHERE ep.ProjectID IS NULL;
```

This query retrieves the names of employees who are not assigned to any projects. It uses a LEFT JOIN to combine the Employees table with the Employees_Projects table on the EmployeeID column. The WHERE ep.ProjectID IS NULL condition filters the results to include only those employees who do not have a corresponding ProjectID in the Employees_Projects table. See Table 4-18.

Table 4-18. *The Result of the Sixth Query*

EmployeeName
John Smith
Jane Dylan
Emily Davis
David White

By using these different types of SQL JOINs, Piper answered a variety of questions about the company, such as manager-report relationships, department-employee associations, project assignments, and identifying non-project employees. Considering the vastness of the data and the many dimensions that existed in the company's data, it would be impossible to answer these questions without using these JOINs.

Handling NULL Values in JOINs

Handling NULL values in SQL joins can be very important to maintain the integrity and accuracy of query results. When dealing with NULL values, it is essential to understand how the different JOIN types behave and the appropriate strategies to apply. This section examines the behavior of various JOINs with NULL values and the way that you can handle them effectively.

NULL Behavior in SQL JOIN

In SQL, NULL indicates the absence of a known value in a database field. It's distinct from empty strings, zeros, or other variables. NULL values often signify missing data or data that is not applicable. However, NULL can behave differently depending on the context:

- **Comparisons**: Comparing a column with NULL usually results in NULL itself, as the exact relationship, for instance equal to, is unknown.

- **Operations**: Mathematical operations involving NULL typically return NULL, as the calculation cannot be performed without a definite value.

- **Aggregate functions**: Some functions like SUM or AVG ignore NULL values during calculation, while COUNT(*) counts all rows, including those with NULL.

- **JOINs**: Different join types generate NULL values differently, affecting which rows appear in the final result set.

Understanding these behaviors is crucial for writing accurate and efficient SQL queries that effectively deal with missing information. The third story goes into more detail about how to handle NULLs when using different types of JOINs.

The Third Story: Hospital Management

Aletta intends to analyze the management data of the hospital where she works. The hospital management system has a database with tables containing data related to doctors, patients, appointments, and departments. These tables contain the data to effectively manage relationships and programs and ensure that every patient receives the care they need. Table 4-19 lists the doctors; Table 4-20 lists the patients; Table 4-21 lists the appointments; and Table 4-22 lists the departments.

Table 4-19. *The Doctors*

doctor_id	Name	department_id
1	Dr. Alice Smith	1
2	Dr. Bob Johnson	2
3	Dr. Charlie Lee	1
4	Dr. Dana White	3
5	Dr. Eve Black	2

Table 4-20. *The Patients*

patient_id	Name	primary_doctor_id
101	John Doe	1
102	Jane Roe	2
103	Jim Beam	NULL
104	Jack Daniels	4
105	Jill Hill	1

Table 4-21. *The Appointments*

appointment_id	patient_id	doctor_id	appointment_date
1001	101	1	2024-07-01
1002	102	2	2024-07-02
1003	103	3	2024-07-03
1004	105	5	2024-07-04
1005	101	4	2024-07-05

Table 4-22. *The Departments*

department_id	name
1	Cardiology
2	Neurology
3	Orthopedics
4	Pediatrics
5	Dermatology

Aletta would like to:

1. Generate a report that lists all appointments with details about the doctor and patient for each appointment.

2. Create a list of all patients and their primary doctor, if any.

3. Identify all doctors and list their patients (primary care only), if any.

Aletta intends to find a list of all appointments along with the details of the doctor and patient for each appointment, but the problem is that she has to exclude any appointments that do not have a matching doctor or patient. The following query can do this for Aletta:

```
SELECT a.appointment_id, p.name AS patient_name, d.name AS doctor_name,
a.appointment_date
FROM Appointments a
```

```
INNER JOIN Patients p ON a.patient_id = p.patient_id
INNER JOIN Doctors d ON a.doctor_id = d.doctor_id;
```

As mentioned, INNER JOIN removes rows that have no matching values in the joined tables. If patient_id in the Appointments table does not match any patient_id in the Patients table or doctor_id in the Appointments matches doctor_id in the Doctors, those rows will not appear in the result. Since NULLs in JOIN columns result in non-matching rows, they are effectively removed from the result set. There is no need to handle extra NULLs in this query because the INNER JOIN naturally filters them out. Table 4-23 shows all appointments along with the details of the doctor and patient for each appointment.

Table 4-23. *A List of All Appointments*

appointment_id	patient_name	doctor_name	appointment_date
1001	John Doe	Dr. Alice Smith	2024-07-01
1002	Jane Roe	Dr. Bob Johnson	2024-07-02
1003	Jim Beam	Dr. Charlie Lee	2024-07-03
1004	Jill Hill	Dr. Eve Black	2024-07-04
1005	John Doe	Dr. Dana White	2024-07-05

Aletta wants to get a list of all patients along with their primary doctor, if any. To do this, she needs to consider a LEFT JOIN, which includes all rows in the Patients table and matches the rows in the Doctors table. If they do not match, the columns of the doctors table will contain NULL. To handle these NULLs, Aletta can use the COALESCE function replace the NULL values in the Doctors.name column with 'No Primary Doctor Assigned'. This ensures that the result set provides meaningful information even when no primary doctor is assigned to the patient.

```
SELECT p.name AS patient_name, COALESCE(d.name, 'No Primary Doctor
Assigned') AS doctor_name
FROM Patients p
LEFT JOIN Doctors d ON p.primary_doctor_id = d.doctor_id;
```

COALESCE is useful for handling NULL values. For example, when you want to display a default value instead of NULL when a particular column has missing data, COALESCE allows you to do this directly in your SQL query.

Note   COALESCE is a function in SQL that helps you deal with NULL values. It evaluates a list of expressions you provide, one by one, and returns the first expression that is not NULL. COALESCE works by considering a list of values or expressions separated by commas. COALESCE starts checking the expressions from left to right. If the first expression is not NULL, COALESCE returns that value immediately. If the first expression is NULL, COALESCE moves on to the second expression and checks if it's NULL. This process continues until it finds a non-NULL value or reaches the end of the list. If all the expressions in the list are NULL, COALESCE itself returns NULL. COALESCE can take any number of expressions as arguments (more than two). It should be noted that the data types of the expressions in the list need to be compatible. COALESCE will return the data type of the first non-NULL expression.

Table 4-24 shows each patient with their primary doctor's name. For patients without an assigned primary doctor, the query substitutes "No Primary Doctor Assigned".

Table 4-24. *A List of All Patients and Their Primary Doctor*

patient_name	doctor_name
John Doe	Dr. Alice Smith
Jane Roe	Dr. Bob Johnson
Jim Beam	No Primary Doctor Assigned
Jack Daniels	Dr. Dana White
Jill Hill	Dr. Alice Smith

To get a list of all doctors along with the patients they are primary doctors for, Aletta uses RIGHT JOIN to include all rows from the Doctors table and matches rows from the Patients table.

```
SELECT COALESCE(p.name, 'No Patients Assigned') AS patient_name, d.name AS
doctor_name
FROM Patients p
RIGHT JOIN Doctors d ON p.primary_doctor_id = d.doctor_id;
```

In this query, the RIGHT JOIN includes all rows from the Doctors table and matches rows from the Patients table. If there is no match, the columns from the Patients table will contain NULL. To Handle NULLs, the COALESCE function is used to replace NULL values in the Patients.name column with 'No Patients Assigned'. This ensures that the result set shows all doctors, including those who are not the primary doctor for any patient, and provides a meaningful placeholder for the NULL values. Table 4-25 shows each doctor with the names of their primary patients. For doctors without an assigned primary patient, the query substitutes 'No Patients Assigned'.

Table 4-25. *Doctors with the Names of Their Primary Patients*

patient_name	doctor_name
John Doe	Dr. Alice Smith
Jane Roe	Dr. Bob Johnson
Jack Daniels	Dr. Dana White
Jill Hill	Dr. Alice Smith
No Patients Assigned	Dr. Eve Black
No Patients Assigned	Dr. Charlie Lee

NULL-Safe Equal Operator

It should be noted that some databases, like in MySQL, offer a NULL-safe equal operator. This operator <=> behaves like the standard =, but with special handling for NULL values. If both operands are NULL, it returns 1 (true). If only one operand is NULL, it returns 0 (false). This simplifies comparisons involving NULLs, avoiding the need for complex logic or additional checks for missing data. It improves code readability and ensures your queries function as expected when encountering NULL values.

PostgreSQL doesn't have a NULL-safe equal operator like <=>. However, PostgreSQL offers alternative ways to achieve NULL-safe comparisons:

- IS NULL **and** IS NOT NULL: These operators explicitly check for the presence or absence of NULL values. You can use them in your comparisons, like so:

```
SELECT *
FROM table
WHERE column1 IS NULL OR column2 = value
```

- **The** = **operator**: While the standard = operator might return NULL when comparing with NULL, you can use PostgreSQL's NULL-safe comparison behavior with the IS DISTINCT FROM operator.

```
SELECT *
FROM table
WHERE column1 IS DISTINCT FROM column2
```

This acts like = for non-NULL values, but returns true when both operands are NULL and false when only one is NULL.

Both approaches effectively handle NULL-safe comparisons in PostgreSQL. While PostgreSQL doesn't have a dedicated NULL-safe operator, its NULL-safe comparison behavior with IS DISTINCT FROM offers similar benefits for writing cleaner and more readable JOIN conditions when dealing with NULL values. Code using NULL-safe operators is easier to understand. Without these operators, you might need nested checks (IS NULL or COALESCE) within your JOIN conditions to handle NULL values. This can make the query logic convoluted and harder to maintain.

Summary

This chapter described an essential tool for combining data from multiple tables, called SQL JOIN, which can enable complex and meaningful data analysis. Understanding the different types of JOINs and their applications is crucial for performing advanced data queries effectively.

Key Points

- INNER JOIN: Combines rows from two or more tables based on a related column, returning only matching rows.

- LEFT OUTER JOIN: Retrieves all rows from the left table and the matched rows from the right table, filling in NULLs for unmatched rows.

- RIGHT OUTER JOIN: Retrieves all rows from the right table and the matched rows from the left table, filling in NULLs for unmatched rows.

- FULL OUTER JOIN: Returns all rows when there is a match in either left or right table, filling in NULLs where there are no matches.

- SELF JOIN: A table is joined with itself to compare rows within the same table.

- CROSS JOIN: Produces a Cartesian product, pairing each row from one table with all rows in another table, used rarely for specific purposes.

- **Handling** NULL **values**: Properly handle NULL values, resulting from outer joins to avoid misleading data.

Key Takeaways

- **Types of** JOINs: Explore INNER, LEFT, RIGHT, and FULL JOINs to combine data from multiple tables based on matching criteria.

- **JOIN conditions**: Define the rules for matching rows between tables using the ON clause and comparison operators.

- **Handling** NULLs: Understand how different JOIN types handle missing data (NULL values) in the results.

Looking Ahead

The next chapter, "Aggregating Acts," explores aggregation functions and GROUP BY, which are fundamental for summarizing and analyzing large datasets. Mastery of this operation will enable you to extract meaningful insights from your data, uncovering trends, patterns, and overall statistics.

Test Your Skills

A database has three tables: customers (CustomerID, Name, City), orders (OrderID, CustomerID, ProductID, OrderDate), and products (ProductID, Name, Description, Price, Color).

1. Write an SQL query using an appropriate JOIN to retrieve the names and cities of all customers who placed an order in July.

2. The products table has a column called Color that might contain NULL values. Write a query that retrieves all products along with their color information. However, for products with a NULL color value, display "Unknown" instead.

3. Write a query to find all pairs of customers who live in the same city. Ensure that each pair is listed only once.

CHAPTER 5

Aggregating Acts

An aggregate act in SQL is the process of applying aggregate functions to grouped data subsets from a data table. To accomplish this, aggregate functions, such as COUNT, SUM, AVG, MIN, and MAX, are combined with the GROUP BY clause to calculate specific groups of data. As a result of these actions, SQL provides valuable insights for data analysis and decision-making by extracting meaningful summaries and statistical information.

Introduction to GROUP BY

In SQL, GROUP BY allows identical data to be grouped together. GROUP BY clauses in SQL allow us to group rows according to one or more columns. Once grouped, aggregate functions can be applied to calculate summary statistics for each group. The GROUP BY clause syntax is as follows:

```
SELECT first_column, aggregate_function(second_column)
FROM table_name
WHERE condition
GROUP BY first_column;
```

Here, the first_column is the column used to group the data, an aggregate_function is a function that calculates a single value from a set of values like COUNT, SUM, AVG, MIN, and MAX, the second_column is the column on which the aggregate function is applied, the table_name refers to the table you are querying, and the WHERE condition filters the data before grouping it.

The GROUP BY clause can be used to divide the result set into subgroups based on the column(s) specified in the clause. There are multiple groups formed by each unique value in the grouping column. Each group is given a single value by applying an aggregate function. The final result contains a row for each group along with the aggregate values.

© Hamed Tabrizchi 2025
H. Tabrizchi, *Narrative SQL*, https://doi.org/10.1007/979-8-8688-1560-7_5

The GROUP BY clause is essential for data aggregation, as it allows you to summarize data, analyze trends, create reports, and improve performance. GROUP BY can summarize data by computing totals, averages, minimums, and maximums. Within different categories of data, GROUP BY can be used to identify patterns and trends. A report can be created by GROUP BY using summarized information for various groups. It is often possible to improve the performance of queries using GROUP BY by reducing the amount of data returned. This chapter elaborates on each of these aspects in more detail. It explores how to compute totals, averages, minimums, and maximums using GROUP BY. Additionally, it explains how to combine aggregations and window functions, use nested GROUP BY for more granular analysis, and simplify complex queries with common table expressions.

Essential Aggregation Functions

In SQL, aggregate functions are powerful tools for calculating data. These functions allow you to calculate meaningful statistics from large datasets, which is essential for the analysis and reporting of data. Table 5-1 lists the most commonly used aggregation functions in SQL.

Table 5-1. *Commonly Used Aggregation Functions*

Function	Purpose	Syntax
COUNT()	Counts rows or non-NULL values	COUNT(column_name) or COUNT(*)
SUM()	Calculates the sum of a numeric column	SUM(column_name)
AVG()	Computes the average of a numeric column	AVG(column_name)
MIN()	Finds the smallest value in a column	MIN(column_name)
MAX()	Finds the largest value in a column	MAX(column_name)
STDDEV()	Calculates the standard deviation of a numeric column	STDDEV(column_name)

(continued)

Table 5-1. (*continued*)

Function	Purpose	Syntax
VARIANCE()	Computes the variance of a numeric column	VARIANCE(column_name)
MEDIAN()	Returns the median value of a numeric column	MEDIAN(column_name)
PERCENTILE_CONT()	Calculates a percentile using continuous distribution	PERCENTILE_CONT(percentile) WITHIN GROUP (ORDER BY column_name)
PERCENTILE_DISC()	Calculates a percentile using discrete distribution	PERCENTILE_DISC(percentile) WITHIN GROUP (ORDER BY column_name)
STRING_AGG()	Concatenates values into a single string	STRING_AGG(column_name, delimiter)
ARRAY_AGG()	Aggregates values into an array	ARRAY_AGG(column_name)
MODE()	Finds the most frequent value in a column	MODE()
BIT_AND()	Computes the bitwise AND of all non-NULL input values	BIT_AND(column_name)
BIT_OR()	Computes the bitwise OR of all non-NULL input values	BIT_OR(column_name)

A wide range of data summarization tasks can be accomplished with these aggregation functions. In addition to covering the most common statistical analysis, data reporting, and aggregation needs, they are also fundamental tools for anyone using PostgreSQL. To begin the journey of effective data grouping, meaningful summaries, and valuable insight extraction, you use these functions along with GROUP BY.

The First Story: A Busy Gym in a Bustling City

This story revolves around a busy gym in a bustling city that has been operating successfully for many years. Jack, the gym data analyst, helps the management team make data-driven decisions to optimize management policies, improve customer satisfaction, and boost profitability. The gym collects a variety of data points daily, including information on memberships, classes, personal training sessions, and overall customer activity.

The gym has several types of memberships, including Basic, Standard, and Premium. It offers various fitness classes such as Yoga, Pilates, and Spinning, and it also has a team of personal trainers who conduct one-on-one sessions. The gym tracks all this data in a relational database. Jack is required to work with the following data tables.

The first table is the Members table and it contains information about gym members, including their membership type. An illustration of this table can be found in Table 5-2.

Table 5-2. *The Members Table*

Member_ID	First_Name	Last_Name	Membership_Type	Join_Date	Date_of_Birth	Phone_Number
1	John	Doe	Premium	2023-01-10	1985-02-20	555-1234
2	Jane	Smith	Standard	2022-08-15	1990-05-15	555-5678
3	Emily	Johnson	Basic	2023-03-20	1992-09-05	555-8765
4	Michael	Brown	Premium	2021-12-10	1988-11-23	555-4321
5	Sarah	Davis	Standard	2023-06-01	1987-07-30	555-6543

The second table is the Classes table, which contains data about the different classes offered at the gym. The table values can be found in Table 5-3.

Table 5-3. *The Classes Table*

Class_ID	Class_Name	Instructor	Class_Type	Schedule	Max_Capacity
1	Yoga	Alice Green	Fitness	Mon 9:00 AM	20
2	Pilates	Bob White	Fitness	Wed 7:00 PM	15
3	Spinning	Carol Black	Cardio	Fri 6:00 PM	25
4	Zumba	Dave Brown	Cardio	Sat 10:00 AM	30
5	HIIT	Eve Silver	Strength	Tue 8:00 AM	20

As shown in Table 5-4, the Attendance table keeps track of which classes members attended and on what dates.

Table 5-4. *The Attendance Table*

Attendance_ID	Member_ID	Class_ID	Attendance_Date
1	1	3	2023-08-18
2	2	1	2023-08-19
3	1	4	2023-08-20
4	3	2	2023-08-20
5	4	5	2023-08-21

The Personal_Training_Sessions table records details of personal training sessions, including the trainer, member, session date, and duration, as shown in Table 5-5.

Table 5-5. *The Personal_Training_Sessions Table*

Session_ID	Trainer_Name	Member_ID	Session_Date	Duration_mins
1	Tom Harris	1	2023-08-17	60
2	Nina Jordan	2	2023-08-18	45
3	Mike Scott	3	2023-08-19	30
4	Lisa White	4	2023-08-19	60
5	Tom Harris	5	2023-08-20	45

The last table (see Table 5-6) is `Payments`, which logs all payments made by members, including membership fees, class fees, and personal training fees.

Table 5-6. *The Payments Table*

Payment_ID	Member_ID	Payment_Date	Amount	Payment_Type
1	1	2023-08-01	50	Membership Fee
2	2	2023-08-01	40	Membership Fee
3	1	2023-08-18	20	Class Fee (Spinning)
4	3	2023-08-19	60	Personal Training Fee
5	4	2023-08-20	50	Class Fee (HIIT)

Jack wants answers to the following questions.

1. For each membership type, what is the average age of the members?

2. How many personal training sessions were conducted by each trainer, and what was the total duration of those sessions?

3. What is the total revenue generated by each type of payment? A membership fee, a class fee, and a personal training fee are all types of payments.

4. How many members attended each type of class, and what is the total number of attendance for each class?

5. For each membership type, how many members have attended at least one class, and what is the average number of classes attended per member?

The following query can be used to calculate the average age of members in different membership types:

```
SELECT  Membership_Type, CAST(AVG(DATE_PART('year', AGE(CURRENT_DATE, Date_
of_Birth))) AS INTEGER) AS Avg_Age
FROM Members
GROUP BY Membership_Type;
```

In this query, the AGE(CURRENT_DATE, Date_of_Birth) function calculates the difference between the current date (CURRENT_DATE) and a person's date of birth (Date_of_Birth). The result is a time interval representing a person's age in years, months, and days. Using DATE_PART('year', AGE(...)), you can access the number of years from the interval returned by AGE() while ignoring months and days. This provides the person's age in complete years. AVG(DATE_PART(...)) is an aggregate function that calculates the average age in years across all rows. CAST(... AS INTEGER) is used because the result of AVG() could have decimal places, so the CAST(... AS INTEGER) converts the result to an integer. This discards any decimal part and provides a whole number for ease of reading. By understanding the typical age profile of members in each age bracket, the gym can offer customized services and marketing. It should be noted that the DATE_PART and AGE functions are standard PostgreSQL functions for working with dates. The use of CURRENT_DATE and CAST are also standard.

Note   The AGE(timestamp1, timestamp2) function calculates the difference between two dates or timestamps and returns the result as an interval(years, months, days, etc.). It should be noted that if only one argument is given in AGE(timestamp2), PostgreSQL assumes CURRENT_DATE as the recent date. For instance, SELECT AGE('2024-01-01', '1990-01-01') returns the time interval from January 1, 1990 to January 1, 2024, showing the age as something like 34 years 0 months 0 days. Also, the DATE_PART() function extracts a specific part (like a year, month, or day) from a date or interval, part is part of the date or interval you want to extract, such as the year, month, or day. date is the date or interval from which you want to extract the value.For instance, SELECT DATE_PART('year', '2024-09-22') returns 2024 as it extracts the year from the date.

Table 5-7 illustrates the calculation of average age based on the membership type.

Table 5-7. *The Average Age Based on Membership Type*

Membership_Type	Avg_Age
Premium	38
Standard	36
Basic	32

Through the following query, you can discover how many sessions a personal trainer has and how much time they spend on each session, which are important metrics to evaluate a trainer's workload and performance.

```
SELECT Trainer_Name,  COUNT(Session_ID) AS Total_Sessions, SUM(Duration_
mins) AS Total_Duration
FROM  Personal_Training_Sessions
GROUP BY Trainer_Name;
```

The query provides insight into the performance of each personal trainer by calculating the total number of training sessions they have conducted and their cumulative duration. By using the COUNT and SUM functions, you can count the number of sessions held by each trainer and calculate the total minutes spent in training sessions. When Trainer_Name is grouped, each trainer's calculations are separate. As a result of this information, trainer workloads can be managed and trainer contributions assessed in the gym. Table 5-8 shows the number and duration of personal training sessions for each trainer.

Table 5-8. *Training Sessions Per Trainer and Their Duration*

Trainer_Name	Total_Sessions	Total_Duration
Tom Harris	2	105
Nina Jordan	1	45
Mike Scott	1	30
Lisa White	1	60

The following query can help identify which services contribute most to its income by understanding its revenue breakdown by payment type:

```
SELECT Payment_Type, SUM(Amount) AS Total_Revenue
FROM Payments
GROUP BY Payment_Type;
```

Each type of payment is calculated separately in this query. The SUM function aggregates the payment amounts, grouped by Payment_Type, including membership fees, class fees, and personal training fees. As a result of this analysis, as shown in Table 5-9, the gym's management can make informed decisions about pricing strategies, promotions, and resource allocation based on its major sources of income.

Table 5-9. *The Gym's Major Sources of Income*

Payment_Type	Total_Revenue
Membership Fee	90.00
Class Fee (Spinning)	20.00
Personal Training Fee	60.00
Class Fee (HIIT)	50.00

To determine how many members attend each class at the gym, the following query can be used:

```
SELECT  C.Class_Name, COUNT(DISTINCT A.Member_ID) AS Number_of_Members,
COUNT(A.Attendance_ID) AS Total_Attendances
FROM  Attendance A
JOIN Classes C ON A.Class_ID = C.Class_ID
GROUP BY C.Class_Name;
```

This query provides two key metrics for each class offered at the gym. These are the number of distinct members who attended at least one session, which is calculated by using COUNT(DISTINCT A.Member_ID)) and the total number of attendances, which is calculated by using COUNT(A.Attendance_ID). To match attendance records with class names, it joins the Attendance table with the Classes table. The analysis is performed separately for each class when classes are grouped by Class_Name. As shown in Table 5-10, gym managers can take advantage of these insights to optimize class schedules, allocate resources more efficiently, and even introduce new classes based on member preferences.

Table 5-10. *Attendance Totals and Membership Numbers for Each Class*

Class_Name	Number_of_Members	Total_Attendances
Spinning	1	1
Yoga	1	1
Zumba	1	1
Pilates	1	1
HIIT	1	1

The following nested query can be used to find, for each membership type, the number of members who attended at least one class, as well as the average number of classes attended per member.

```
SELECT M.Membership_Type, COUNT(DISTINCT M.Member_ID) AS Members_Attended,
CAST(AVG(Sub.Class_Attended) AS DECIMAL(10, 2)) AS Avg_Classes_Per_Member
FROM Members M
JOIN
    (SELECT Member_ID, COUNT(Attendance_ID) AS Class_Attended
     FROM Attendance
     GROUP BY Member_ID) Sub
ON M.Member_ID = Sub.Member_ID
GROUP BY M.Membership_Type;
```

This nested query aims to analyze class attendance to determine membership engagement levels. As a first step, a subquery finds the number of classes each member attended. In the main query, the result is then joined to the Members table, and the data is grouped by Membership_Type. This calculates the average number of classes attended by members of each membership type.

As mentioned earlier, nested queries are queries embedded within other queries. In this query, the subquery calculates how many classes each member has attended by selecting Member_ID and counting the occurrences of Attendance_ID in the Attendance table. After that, the results are grouped by Member_ID, giving a list of members and their respective class counts. This subquery is then given an alias Sub and used in the main query to join the Members table. As a result, the main query can access membership data and aggregate it to derive insights, such as how many members of each membership type attended at least one class and how many classes they attended on average. Table 5-11 shows member engagement, which is useful for understanding how active each membership type is. These insights can be used to customize the gym's offerings or create membership promotion offers.

Table 5-11. *Analysis of Class Attendance Based on Type of Membership*

Membership_Type	Members_Attended	Avg_Classes_Per_Member
Premium	2	1.5
Standard	1	1
Basic	1	1

By using SQL's GROUP BY and aggregation capabilities, these queries reveal crucial information about the gym statistics, participation, and revenue. The next section discusses a more advanced topic.

Advanced Aggregation Techniques: Multi-step Calculations

This section discusses a more advanced topic, called *multi-step calculations*.

Multi-step Calculations: The Basics

For complex data analysis tasks requiring multiple calculation steps, subqueries and nested aggregations can be applied together. Subqueries can be used in various places within the main query, such as in a SELECT, FROM, or WHERE clause. Subqueries typically return temporary results that the main query can reference. In essence, you can break complex operations down into smaller, more manageable queries, which can be used to perform specific calculations for the main query. To achieve nested aggregation, subqueries are commonly used. Performing multiple levels of aggregation, namely processing or aggregating the result of one aggregation by another, is called *nested aggregation*. Subqueries are generally aggregated, and then another layer of aggregation is applied. The primary reason for using nested aggregations is to calculate metrics that require multi-step computations, such as group averages, rankings, or aggregated summaries. As a result, you can perform complex aggregations on datasets by making intermediate calculations.

Assuming the previous story scenario, a busy gym in a bustling city, if you want to calculate the average number of classes attended by members, grouped by membership type, you need to break this down into two steps. First, you count the number of classes each member has attended, and then you calculate the average of those counts. The first aggregation is performed with a subquery, counting attendance per member, and the second aggregation, averaging counts across membership types, is performed with the outer query.

```
SELECT M.Membership_Type, AVG(Sub.Class_Attended) AS Avg_Classes_Per_Member
FROM Members M
JOIN (SELECT Member_ID, COUNT(Attendance_ID) AS Class_Attended FROM
Attendance GROUP BY Member_ID) AS Sub
ON M.Member_ID = Sub.Member_ID
GROUP BY M.Membership_Type;
```

Here, the subquery (`SELECT Member_ID, COUNT(Attendance_ID) AS Class_Attended FROM Attendance GROUP BY Member_ID`) calculates the total number of classes each member has attended, and the outer query then performs a second aggregation, calculating the average (`AVG(Sub.Class_Attended)`) number of classes attended by members within each membership type, grouping the data by `Membership_Type`.

The advantages of combining subqueries and nested aggregation are that you get better modularity, flexibility, and complex analysis. Subqueries allow for modular code, where intermediate calculations can be separated from the main logic, making the query more readable and easier to debug. Also, using subqueries for nested aggregation offers flexibility to perform multiple levels of aggregation that are otherwise difficult to achieve in a single pass. Additionally, this leads to the creation of complex analyses like ranking, cumulative sums, and multi-step calculations, where one aggregation depends on another's result.

Using Window Functions for Aggregation

The window function is a powerful SQL tool that allows you to perform calculations across sets of rows related to the current query row. Unlike traditional aggregation functions, such as SUM(), COUNT(), or AVG(), used with GROUP BY, window functions do not collapse rows into a single result. Instead, they provide a calculated value while maintaining the row structure.

Window Function Definition

Window functions operate on a set of rows called a "window," which is defined by the OVER() clause. The OVER() clause specifies how rows should be grouped or partitioned and evaluated. By applying the window function to each row within the window, the result set is not collapsed. Window functions involve several key concepts that should be noted. Partitioning is the first concept. It is possible to divide the data into groups or "partitions" based on one or more columns by using the PARTITION BY clause. The second concept is ordering. With the ORDER BY clause, you can specify the order of rows within each partition. The last concept is framing. If you want to control which rows appear in the window, you can use the ROWS or RANGE options, but this is less commonly required for simple aggregations.

A window function has the following basic syntax:

```
<window_function> OVER ( [PARTITION BY <column>] [ORDER BY <column>] )
```

Here, <window_function>() can be any SQL aggregate or ranking function, such as SUM(), AVG(), COUNT(), ROW_NUMBER(), RANK(), and so on. The OVER() clause specifies the "window" or the set of rows that the function should operate over. PARTITION BY divides the result set into partitions, similar to GROUP BY, and ORDER BY defines the order of rows within each partition.

Also, the basic structure of a window function using the ROWS or RANGE options is as follows:

```
<window_function> OVER (
    [PARTITION BY <column_name>]
    ORDER BY <column_name>
    {ROWS | RANGE} BETWEEN <frame_start> AND <frame_end>
)
```

Here, ROWS defines the window in terms of physical rows, and RANGE defines the window in terms of logical values (ranges of values). It is important to note that ROWS operates on a specific number of rows relative to the current row, and RANGE operates on a range of values (not necessarily consecutive rows).

Note   The ORDER BY clause in SQL is used to sort the result set of a query by one or more columns. This clause allows you to display the data alphabetically, numerically, or by date. This clause is introduced in the next chapter with more examples and stories. Briefly, it can be said that the basic syntax of ORDER BY is as follows:

```
SELECT column_1, column_2
FROM table_name
ORDER BY column1 [ASC];
```

ORDER BY column1 specifies the column you want to sort by, and ASC means ascending the data from smallest to largest. In the absence of a specification, ASC order is the default. As well, DESC means sorting the data in descending order, from largest to smallest.

The Second Story: Speedy Motors Company

At a fast-growing automobile manufacturer, management is excited to gather insights from a data table that contains information about sales, customers, and vehicles. Their company hired Anna, a data analyst, to help them make sense of the data and provide meaningful insights to drive business decisions. The company collects data on car sales, customer demographics, pricing, and production. Anna needs to perform the following advanced data analysis in order to answer their key business questions. Anna is working with the Car Sales table, shown in Table 5-12, which contains records of car sales, including the region, model name, sale date, and sales amount.

Table 5-12. *The Car Sales Table*

Sale_ID	Model_Name	Region	Sale_Date	Sales_Amount
1	Speedster	North	2024-01-01	25000
2	Cruiser	South	2024-01-03	20000
3	Speedster	North	2024-01-05	27000
4	Zoomer	East	2024-01-07	30000
5	Speedster	West	2024-01-10	28000
6	Cruiser	South	2024-01-12	22000
7	Zoomer	North	2024-01-15	26000
8	Speedster	South	2024-01-17	29000

Anna needs answers to the following questions.

1. How can they identify the order in which cars were sold in each region?

2. How can they rank car models by their total sales in each region?

3. How can they rank car models without skipping ranks when sales tie? (A sales tie occurs when two or more car models have the same number of sales within a specific region.)

4. How can they calculate the cumulative sales for each car model over time?

5. How can they calculate the moving average of sales over the last three sales for each model?

6. How can they calculate the total sales per region and rank the car models within each region?

To find the answer to the first question, Anna uses ROW_NUMBER() to assign a unique rank to each sale in every region, ordered by sale date, to understand the sequence of sales.

```
SELECT Sale_ID, Region, Model_Name,
       ROW_NUMBER() OVER (PARTITION BY Region ORDER BY Sale_Date) AS
       Sale_Rank
FROM Car_Sales;
```

The query is used to rank each sale of cars within a region, according to the sale date. The query retrieves four columns from the Car_Sales table: Sale_ID, Region, Model_Name, and a newly computed column called Sale_Rank, which is created using the ROW_NUMBER() window function. As the first step, SELECT Sale_ID, Region, Model_Name retrieves the unique sale identifier (Sale_ID), the region in which the car was sold (Region), and the model of the car (Model_Name) from the Car_Sales table. Then, ROW_NUMBER() OVER (PARTITION BY Region ORDER BY Sale_Date) uses the ROW_NUMBER() window function to assign a sequential row number (starting from 1) to each row within a specific partition. Anna uses ROW_NUMBER() to assign a unique rank to each sale in every region, ordered by sale date, to understand the sequence of sales. The partitioning is done by the Region column, meaning row numbers will be generated separately for each region. Within each region, the rows are ordered by Sale_Date in ascending order. This means the earliest sale will have row number 1, the second earliest sale will have 2, and so on. The calculated row number is given the alias Sale_Rank in the result set, AS Sale_Rank, representing the rank or sequence of each sale in its respective region.

Note   PARTITION BY is a clause used in window functions in SQL to divide a result set into smaller groups or partitions. A key difference between PARTITION BY and GROUP BY is that PARTITION BY doesn't collapse the rows; it allows calculations to be performed within each partition while keeping all rows visible. PARTITION BY allows window functions to be applied separately to each partition.

Results will include a list of sales ranked by their order of occurrence within the same region. For tracking sales sequences or identifying the first, second, and so on, sale within each region over time, this is particularly useful. Table 5-13 shows the rank of each sale within its region.

Table 5-13. *The Sales Sequence*

Sale_ID	Region	Model_Name	Sale_Rank
1	North	Speedster	1
3	North	Speedster	2
7	North	Zoomer	3
2	South	Cruiser	1
6	South	Cruiser	2
8	South	Speedster	3
4	East	Zoomer	1
5	West	Speedster	1

This query can be used to rank car models in each region based on their total sales:

```
SELECT Model_Name, Region,
      RANK() OVER (PARTITION BY Region ORDER BY SUM(Sales_Amount) DESC) AS
Sales_Rank
FROM Car_Sales
GROUP BY Model_Name, Region;
```

This query ranks car models based on their total sales within each region using a window function with RANK() and PARTITION BY. The query selects two columns, Model_Name and Region. Then, RANK() OVER (PARTITION BY Region ORDER BY SUM(Sales_Amount) DESC) AS Sales_Rank uses the RANK() function to assign a rank to each car model within its region based on total sales. In this query, Anna uses RANK() to rank car models in each region based on their total sales. Here, the models with the same total sales will receive the same rank. PARTITION BY Region splits the data into partitions by region, so that the ranking calculation is applied separately to each region. By using ORDER BY SUM(Sales_Amount) DESC, Anna ranks the car models in each region based on their total sales SUM(Sales_Amount), sorted in descending order. The highest sales get the highest rank. It should be noted that the rank is reset for each region due to partitioning. GROUP BY Model_Name, Region is used because Anna is applying SUM(Sales_Amount). Thus, the query groups the data by Model_Name and Region to compute the total sales for each model in each region. Finally, each car model is ranked within its region based on total sales, with the model having the highest sales ranking first. Table 5-14 shows the rank of car models by total sales in each region.

Table 5-14. *Ranking Car Models by Sales in Each Region*

Model_Name	Region	Sales_Rank
Speedster	North	1
Zoomer	North	2
Cruiser	South	1
Speedster	South	2
Zoomer	East	1
Speedster	West	1

In order to rank car models in order of sales without skipping ranks when there is a tie in sales, the next query is ranked sequentially without gaps. To answer the third question, models should be ranked without gaps when sales figures are tied. In order to avoid gaps when two models have the same sales amount, this query uses DENSE_RANK():

```
SELECT Model_Name, Region,
       DENSE_RANK() OVER (PARTITION BY Region ORDER BY SUM(Sales_Amount)
       DESC) AS Dense_Rank
FROM Car_Sales
GROUP BY Model_Name, Region;
```

This query ranks car models by their total sales within each region using the DENSE_RANK() window function. The query selects two columns: Model_Name and Region. DENSE_RANK() OVER (PARTITION BY Region ORDER BY SUM(Sales_Amount) DESC) AS Dense_Rank, DENSE_RANK() is a window function that assigns a rank to each car model based on its sales amount within a region. DENSE_RANK() ranks models without gaps when sales figures are tied. PARTITION BY Region partitions data by Region, meaning the ranking is calculated separately for each region. ORDER BY SUM(Sales_Amount) DESC ranks the car models based on their total sales (SUM(Sales_Amount)), with the models having the highest sales ranked first.

> **Note**   Unlike RANK(), which skips ranks if ties occur, DENSE_RANK() does not skip numbers. If two models have the same sales, they will share the same rank, and the next model will be ranked right after (without gaps). A final step is to use GROUP BY Model_Name, Region to calculate the total sales for each model based on SUM(Sales_Amount).

As a result of the query, the ranking of car models is dense in each region based on their total sales, ensuring that no gaps exist in the rank numbers, even when there are ties among models. Table 5-15 ranks car models without skipping ranks, even when there are ties in total sales.

Table 5-15. *Dense Ranking of Car Models by Sales*

Model_Name	Region	Dense_Rank
Speedster	North	1
Zoomer	North	2
Cruiser	South	1
Speedster	South	2
Zoomer	East	1
Speedster	West	1

The following query calculates the cumulative sales over time for each car model:

```
SELECT Sale_ID, Model_Name, Sale_Date,
    SUM(Sales_Amount) OVER (PARTITION BY Model_Name ORDER BY Sale_Date)
    AS Cumulative_Sales
FROM Car_Sales;
```

Anna uses the SUM() function with the OVER() clause to calculate a running total of sales for each car model, ordered by sale date. A cumulative sales calculation for each car model is performed from the Car_Sales table using this query. SUM(Sales_Amount) OVER (PARTITION BY Model_Name ORDER BY Sale_Date) computes a running total of sales for each Model_Name, ordered by Sale_Date. The method returns the Sale_ID, Model_Name, Sale_Date, and the cumulative sales amount up until that sale. Table 5-16 shows the cumulative total of sales for each car model over time.

Table 5-16. *Cumulative Sales by Car Model*

Sale_ID	Model_Name	Sale_Date	Cumulative_Sales
1	Speedster	2024-01-01	25000
3	Speedster	2024-01-05	52000
5	Speedster	2024-01-10	80000
8	Speedster	2024-01-17	109000
2	Cruiser	2024-01-03	20000
6	Cruiser	2024-01-12	42000
4	Zoomer	2024-01-07	30000
7	Zoomer	2024-01-15	56000

In order to calculate the moving average of sales for each model over the last three sales, you can use the following query.

```
SELECT Sale_ID, Model_Name, Sale_Date,
       AVG(Sales_Amount) OVER (PARTITION BY Model_Name ORDER BY Sale_Date
       ROWS BETWEEN 2 PRECEDING AND CURRENT ROW) AS Moving_Avg
FROM Car_Sales;
```

Anna uses the AVG() window function to calculate a moving average of the sales amount over the last three sales for each model. With this query, a moving average is calculated for every car model in the Car_Sales table based on its Sales_Amount. To compute the average, the query uses the AVG() window function. PARTITION BY Model_Name groups the data by car model. ORDER BY Sale_Date orders the sales chronologically. ROWS BETWEEN 2 PRECEDING AND CURRENT ROW limits the window to include the current sale and the previous two sales. Moving_Avg is a moving average of sales amounts over time for each model. The window frame determines which rows should be included in the calculation for each row in the result set. Thus, ROWS specifies that the window frame is defined by a specific number of rows relative to the current row (as opposed to a range of values with RANGE). The "BETWEEN 2 PRECEDING AND CURRENT ROW", "2 PRECEDING" includes the current row and the two rows before it in the calculation, and CURRENT ROW refers to the current row itself. Due to this, each row in

the result set will be averaged over the two previous rows (based on the ORDER BY Sale_
Date). When there are fewer than two preceding rows (e.g., for the first or second row), it
will just calculate the average. Table 5-17 shows the moving average of sales amounts for
the last three sales of each car model.

Table 5-17. *Moving Averages of Sales by Car Model*

Sale_ID	Model_Name	Sale_Date	Moving_Avg
1	Speedster	2024-01-01	25000
3	Speedster	2024-01-05	26000
5	Speedster	2024-01-10	26666
8	Speedster	2024-01-17	28000
2	Cruiser	2024-01-03	20000
6	Cruiser	2024-01-12	21000
4	Zoomer	2024-01-07	30000
7	Zoomer	2024-01-15	28000

The following query calculates the total sales per car model within each region, ranks
the models based on their sales within the same region, and provides the total sales for
each region. The inner query first aggregates the total sales for each car model in each
region by summing Sales_Amount from the Car_Sales table. The results are grouped by
Model_Name and Region, which means it calculates the total sales for each model in every
region. The outer query then adds two additional computations: it calculates the total
sales for each region and ranks the models within each region based on their sales.

```
SELECT Model_Name, Region, Total_Sales,
       SUM(Total_Sales) OVER (PARTITION BY Region) AS Total_Sales_Region,
       RANK() OVER (PARTITION BY Region ORDER BY Total_Sales DESC) AS
       Sales_Rank
FROM (
SELECT Model_Name, Region,
       SUM(Sales_Amount) AS Total_Sales
FROM Car_Sales
GROUP BY Model_Name, Region
) AS AggregatedSales;
```

The query starts by aggregating sales in the inner subquery, which calculates the total sales, Total_Sales, for each car model, the Model_Name in each region, and the Region by summing the Sales_Amount and grouping the results by model and region. In the outer query, SUM(Total_Sales) OVER (PARTITION BY Region) is used to calculate the total sales across all models within each region. This provides the total sales for each region, regardless of the model. The RANK() OVER (PARTITION BY Region ORDER BY Total_Sales DESC) then assigns a rank to each model within the region based on its total sales, with the highest sales being ranked first. The query thus provides a view of the total sales for each model, its rank within each region, and the total sales for the region itself. This approach is useful for analyzing car sales performance both by model and region. Table 5-18 shows total sales in each region and ranks the car models based on sales.

Table 5-18. *Total Sales and Ranking by Region*

model_name	Region	total_sales	total_sales_region	sales_rank
Zoomer	East	30000	30000	1
Speedster	North	52000	78000	1
Zoomer	North	26000	78000	2
Cruiser	South	42000	71000	1
Speedster	South	29000	71000	2
Speedster	West	28000	28000	1

To summarize, Anna analyzes Speedy Motors Company's car sales using window functions like ROW_NUMBER(), RANK(), SUM(), and AVG(). The analysis consists of identifying sales sequences, ranking car models by region, calculating cumulative sales, and calculating moving averages for each model. She helps the company make data-driven decisions by combining multiple window functions.

Window Functions vs. Traditional Aggregation

Window functions and traditional aggregation serve different purposes in SQL. As mentioned earlier, traditional aggregation—using GROUP BY with functions like SUM(), COUNT(), AVG(), and so on—combines rows into grouped summaries. A GROUP BY clause, for example, will group data by region and return one row per group if you want the total sales per region. Consequently, row-level data cannot be retained during aggregation. On the other hand, a window function uses OVER() to aggregate rows within a window without collapsing the result set. Each row of data is preserved while running totals, ranks, or moving averages are calculated. For example, SUM(Sales) OVER (PARTITION BY Region ORDER BY Date) calculates cumulative sales per region, but all original rows remain in the output.

The key difference between using window functions and traditional aggregation is that aggregation provides summary data, whereas window functions allow for calculations at the row level based on aggregates. In cases where row-level detail is of importance, window functions are ideal for performing tasks such as ranking, summation, and comparing individual data points against averages across groups, in which the row-level details need to be preserved. See Figure 5-1.

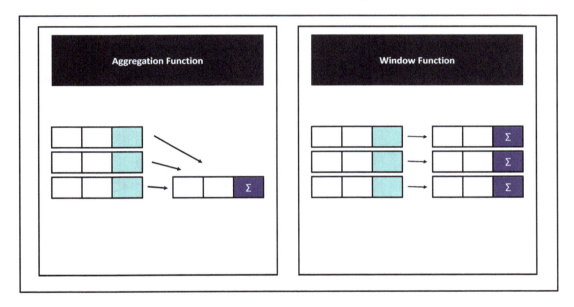

Figure 5-1. *A comparison of window functions and traditional aggregation*

Figure 5-1 illustrates aggregation on the left using GROUP BY, which combines and then hides individual rows. Alternatively, the window function, on the left, can access individual rows and add attributes from those rows. To summarize, traditional aggregation can be used for overall summaries, whereas window functions can be used for more complex, row-aware analysis.

Combining Multiple Aggregation Techniques

The combination of window functions and standard aggregation can lead to more powerful and flexible SQL queries. By combining row-level detail analysis with summary analysis, you can gain more insights than would be possible through either approach alone. To perform complex analyses, you can use window functions, such as ROW_NUMBER() and RANK(), alongside standard aggregate functions.

Combining Window Functions with Standard Aggregation

The following examples demonstrate how to combine window functions and standard aggregation.

The Sales table, shown in Table 5-19, is used for all the queries in this section. It contains data on individual sales, including Customer_ID, Product_ID, Sale_Amount, Sale_Date, and Region.

Table 5-19. *The Sales Table*

Sale_ID	Customer_ID	Product_ID	Sale_Amount	Sale_Date	Region
1	101	P001	500	2024-01-05	North
2	102	P002	300	2024-01-08	South
3	101	P003	200	2024-01-12	North
4	103	P001	900	2024-01-15	East
5	104	P002	600	2024-01-18	West
6	102	P003	150	2024-01-22	South
7	101	P002	700	2024-01-25	North
8	105	P003	250	2024-01-30	West
9	103	P001	400	2024-02-01	East
10	104	P003	1000	2024-02-05	West

Example: Customer Segmentation Based on Purchase Totals

Customers should be ranked based on their total purchases within a specific time frame while still seeing individual sales.

```
SELECT  Customer_ID,
    COUNT(*) OVER (PARTITION BY Customer_ID) AS Total_Sales,
    SUM(Sale_Amount) OVER (PARTITION BY Customer_ID) AS Total_Purchases,
    ROW_NUMBER() OVER (PARTITION BY Customer_ID ORDER BY Sale_Date) AS
    Sale_Rank
FROM Sales;
```

COUNT(*) OVER (PARTITION BY Customer_ID) AS Total_Sales, COUNT(*) is an aggregate function that counts the number of rows. The OVER (PARTITION BY Customer_ID) window function partitions the data by Customer_ID, meaning the count is calculated separately for each customer. The total number of sales made by a customer is shown in this column for each row of sales. The rows are not collapsed, so every row for a customer has the same total sales amount. SUM(Sale_Amount) OVER (PARTITION BY Customer_ID) is a window function that calculates the total purchases for each customer without collapsing the rows, and ROW_NUMBER() OVER (PARTITION BY Customer_ID ORDER BY Sale_Date) assigns a row number to each sale for each customer based on the sale date. In this way, you can see each sale, the total purchase amount for each customer, and the order in which the sales took place.

Table 5-20 shows the number of sales and the total purchase amount for each customer while ranking individual sales by date. The Total_Sales column shows how many sales each customer has made. Total_Purchases displays the total amount the customer has spent. Sale_Rank assigns a number to each sale for a customer, ordered by sale date.

Table 5-20. *Ranking Sales and Calculating Totals per Customer*

Customer_ID	Total_Sales	Total_Purchases	Sale_Rank
101	3	400	1
101	3	1400	2
101	3	1400	3
102	2	450	1
102	2	450	2
103	2	1300	1
103	2	1300	2
104	2	1600	1
104	2	1600	2
105	1	250	1

Nested GROUP BY with Window Functions

For deeper analysis, you can use window functions in conjunction with nested GROUP BY.

Example: Top-N Analysis Within Partitions

The goal here is to determine the three highest-paying customers in each region based on the three most profitable customers.

```
WITH CustomerTotals AS (
    SELECT Customer_ID, Region, SUM(Sale_Amount) AS Total_Purchase
    FROM Sales
    GROUP BY Customer_ID, Region
)
SELECT
    Customer_ID,
    Region,
    Total_Purchase,
```

```
    RANK() OVER (PARTITION BY Region ORDER BY Total_Purchase DESC) AS
    Region_Rank
FROM CustomerTotals
WHERE Customer_ID IN (
    SELECT Customer_ID
FROM (
SELECT
    Customer_ID,
    Region,
    Total_Purchase,
    RANK() OVER (PARTITION BY Region ORDER BY Total_Purchase DESC) AS
    Region_Rank
FROM CustomerTotals ) AS RankedCustomers
WHERE Region_Rank <= 3 );
```

This SQL query uses a *common table expression* (CTE) named CustomerTotals to calculate and organize customer purchase data in a structured manner. The CTE first aggregates the total purchase amount, Total_Purchase, for each customer, Customer_ID, in every region, Region, by totaling Sale_Amount from the Sales table and grouping by both customer and region. This step isolates the aggregation logic, producing a concise result set that contains one row per customer per region, summarizing their total spending.

In the main query, these aggregated data are further processed using the RANK() window function. The function partitions the data by Region and orders the results in descending order of Total_Purchase, assigning a rank to each customer within their respective region. To identify the top performers, the query applies a filter. Only customers whose regional rank is less than or equal to three are selected, effectively limiting the output to the top three spenders in each region. This filtering is done by nesting another query in the ranking results and using a WHERE Region_Rank <= 3 clause. The final SELECT returns Customer_ID, Region, Total_Purchase, and their rank in that region.

The use of a CTE improves the readability and maintainability of the query by clearly separating the total purchase aggregation from the ranking and filtering logic. This approach avoids redundancy by calculating totals once and referencing them multiple times without repeating computations, much like using a temporary, named result set. In complex analytical queries, CTEs simplify debugging, enhance performance through reuse, and make SQL easier to extend or modify in the future.

> **Note**   Common table expressions are temporary result sets that you can
> reference within SELECT, INSERT, UPDATE, or DELETE queries. They are defined
> using the WITH keyword and provide a way to write cleaner, more readable
> queries, especially for complex multi-step operations. The concept of CTEs is
> similar to that of subqueries, but they are much easier to read and reuse. The basic
> syntax of CTEs can be summed up as follows:

```
WITH cte_name AS (
    SELECT columns
    FROM table
    WHERE conditions
)
SELECT * FROM cte_name;
```

Table 5-21 shows the top two customers from each region based on their total
purchases.

Table 5-21.   *Customers in Each Region by Total Purchases*

Customer_ID	Region	Total_Purchase	Region_Rank
101	North	1400	1
102	South	450	1
103	East	1300	1
104	West	1600	1
105	West	250	2

Using Aggregation Functions and ROW_NUMBER() Together

In addition to calculating aggregated metrics like SUM() or COUNT() across partitions, you
can also calculate details and ranks at a row-level.

Example: Count Orders Per Customer and Rank Sales

To calculate the total number of sales for each client, rank each sale, and determine the total amount purchased for each client, you need to calculate the total number of sales.

```
SELECT
    Customer_ID,
    COUNT(*) OVER (PARTITION BY Customer_ID) AS Total_Sales,
    SUM(Sale_Amount) OVER (PARTITION BY Customer_ID) AS Total_Purchases,
    ROW_NUMBER() OVER (PARTITION BY Customer_ID ORDER BY Sale_Date) AS
    Sale_Rank
FROM Sales;
```

COUNT(*) OVER (PARTITION BY Customer_ID) gives the total number of sales for each customer. SUM(Sale_Amount) OVER (PARTITION BY Customer_ID) calculates the total sales amount for each customer. ROW_NUMBER() ranks the sales in order of sale date. Table 5-22 shows the total number of sales and the rank of each sale by customer.

Table 5-22. *Total Sales and Ranking for Each Customer*

Customer_ID	Total_Sales	Total_Purchases	Sale_Rank
101	3	1,400	1
101	3	1,400	2
101	3	1,400	3
102	2	450	1
102	2	450	2
103	2	1,300	1
103	2	1300	2
104	2	1600	1
104	2	1600	2
105	1	250	1

Advanced Query Structures Using Common Table Expressions (CTEs)

In SQL, a common table expression (CTE) represents a temporary result set that can be referred to within a query. CTEs are defined using the WITH keyword. In a nutshell, CTEs create virtual tables that exist only while a query is executed. CTE makes complex queries more readable; it decomposes them into logical, manageable steps. It provides reusable logic that avoids duplication by defining a result set once and reusing it throughout the query. Also, by using CTEs, recursive queries can handle hierarchical and recursive structures. CTEs can also simplify complex multi-step aggregations by breaking them into manageable steps. A CTE allows you to create a temporary result set that can be referenced in the main query. The basic syntax for a CTE is provided here:

```
WITH cte_name AS (
    SELECT column1, column2
    FROM table_name
    WHERE conditions
)
SELECT *
FROM cte_name;
```

In the query, a CTE is defined (cte_name) by using the WITH clause to create a temporary result set from a SELECT statement to be used later. It is possible to refer to this result in the main query (SELECT * FROM cte_name). It simplifies and improves the readability of queries.

Example: Customer Segmentation with Advanced Metrics

Suppose you want to segment customers based on their total purchases into high, medium, and low spenders. The total amount they spent can be calculated first, and then they can be assigned to segments accordingly.

```
WITH CustomerPurchases AS (
    SELECT Customer_ID, SUM(Sale_Amount) AS Total_Purchase
    FROM Sales
    GROUP BY Customer_ID
),
```

```
CustomerSegments AS (
    SELECT Customer_ID, Total_Purchase,
            CASE
                WHEN Total_Purchase >= 1000 THEN 'High Spender'
                WHEN Total_Purchase BETWEEN 500 AND 999 THEN 'Medium
                Spender'
                ELSE 'Low Spender'
            END AS Segment
    FROM CustomerPurchases
)
SELECT * FROM CustomerSegments;
```

The first CTE, CustomerPurchases, calculate the total purchase amount for each customer. The second CTE, CustomerSegments, assign customers into segments (High, Medium, and Low) based on their total purchase. This method makes it easier to break complex queries into manageable steps and improves readability. Table 5-23 shows the customer segmentation based on total purchases.

Table 5-23. *Customer Segments Based on Total Purchases*

Customer_ID	Total_Purchase	Segment
101	1400	High Spender
102	450	Low Spender
103	1300	High Spender
104	1600	High Spender
105	250	Low Spender

Top-N Analysis with CTEs and ROW_NUMBER()

The top-N analysis can be made more complex by combining CTEs and window functions.

Example: Finding the Top Two Highest-Paying Customers Overall

```
WITH RankedCustomers AS (
    SELECT Customer_ID, SUM(Sale_Amount) AS Total_Purchase,
           ROW_NUMBER() OVER (ORDER BY SUM(Sale_Amount) DESC) AS
Purchase_Rank
    FROM Sales
    GROUP BY Customer_ID
)
SELECT * FROM RankedCustomers
WHERE Purchase_Rank < 3;
```

ROW_NUMBER() assigns a rank to each customer based on their total purchases. CTE is used to calculate the total purchase for each customer, rank them, and then filter out the top three. Table 5-24 shows the top two customers with the highest total purchases.

Table 5-24. *Top Two Customers Overall Based on Total Purchases*

Customer_ID	Total_Purchase	Purchase_Rank
104	1600	1
101	1400	2

Essential Window Functions for Data Analysis

Table 5-25 summarizes the most useful window functions in PostgreSQL, along with a brief description and an example of how they are used.

Table 5-25. *The Summarization of the Most Useful Window Functions*

Window Function	Description	Example
ROW_ NUMBER()	Assigns a unique sequential number to each row within a partition, starting from 1.	SELECT ROW_NUMBER() OVER (PARTITION BY Region ORDER BY Sale_Date) AS Row_Num FROM Car_Sales;
RANK()	Ranks each row within a partition, skipping the next rank if there are ties.	SELECT RANK() OVER (PARTITION BY Region ORDER BY Sales_Amount DESC) AS Sales_Rank FROM Car_Sales;
DENSE_ RANK()	Similar to RANK(), but no gaps are left in the ranking numbers after ties.	SELECT DENSE_RANK() OVER (PARTITION BY Region ORDER BY Sales_Amount DESC) AS Dense_Sales_Rank FROM Car_Sales;
NTILE(N)	Divides the result set into N roughly equal groups and assigns a group number to each row.	SELECT NTILE(4) OVER (ORDER BY Sales_Amount DESC) AS Quartile FROM Car_Sales;
SUM()	Calculates the cumulative or running sum of a column over a specified window.	SELECT SUM(Sales_Amount) OVER (PARTITION BY Model_Name ORDER BY Sale_Date) AS Running_Total FROM Car_Sales;
AVG()	Calculates the average of a column over a specified window.	SELECT AVG(Sales_Amount) OVER (PARTITION BY Region ORDER BY Sale_Date) AS Avg_Sales FROM Car_Sales;
MAX()	Returns the maximum value within a partition or over the entire window.	SELECT MAX(Sales_Amount) OVER (PARTITION BY Region) AS Max_Sale FROM Car_Sales;
MIN()	Returns the minimum value within a partition or over the entire window.	SELECT MIN(Sales_Amount) OVER (PARTITION BY Region) AS Min_Sale FROM Car_Sales;

(*continued*)

Table 5-25. (*continued*)

Window Function	Description	Example
COUNT()	Returns the count of rows within a partition or over the entire window.	SELECT COUNT(*) OVER (PARTITION BY Model_Name) AS Sale_Count FROM Car_Sales;
FIRST_VALUE()	Returns the first value within the window frame for each row.	SELECT FIRST_VALUE(Sales_Amount) OVER (PARTITION BY Region ORDER BY Sale_Date) AS First_Sale FROM Car_Sales;
LAST_VALUE()	Returns the last value within the window frame for each row.	SELECT LAST_VALUE(Sales_Amount) OVER (PARTITION BY Region ORDER BY Sale_Date) AS Last_Sale FROM Car_Sales;
LAG()	Returns the value from the previous row in the window frame. Useful for comparing a current row with a previous row.	SELECT LAG(Sales_Amount, 1) OVER (PARTITION BY Model_Name ORDER BY Sale_Date) AS Previous_Sale FROM Car_Sales;
LEAD()	Returns the value from the next row in the window frame. Useful for comparing a current row with the next row.	SELECT LEAD(Sales_Amount, 1) OVER (PARTITION BY Model_Name ORDER BY Sale_Date) AS Next_Sale FROM Car_Sales;

Summary

This chapter explores the process of aggregating data in SQL, which is crucial for summarizing and analyzing data. By understanding how to use aggregate functions with the GROUP BY clause, you can efficiently compute totals, averages, and other statistical values for specific data groups. The advanced topics discussed in this chapter include combining aggregation with window functions, using nested GROUP BY for deeper analysis, and simplifying complex queries with common table expressions.

Key Points

- **Aggregate functions and the** GROUP BY **clause**: Use them to efficiently compute statistics for specific groups of data.

- **Combining aggregations and window functions**: Use SUM(), COUNT(), and ROW_NUMBER(), or RANK() together to calculate both detailed row-level metrics and aggregated summaries.

- **Nested GROUP BY and window functions**: Use GROUP BY to aggregate data first and then apply window functions for deeper analysis, like top-N analysis within partitions.

- **CTEs (common table expressions)**: Simplify multi-step aggregation queries by breaking them into separate logical steps, making your queries easier to read and maintain.

Key Takeaways

- **Aggregate functions**: The COUNT, SUM, AVG, MIN, and MAX functions are useful when working with grouped data.

- **GROUP BY clause**: Organize data effectively for aggregation and summarize results in a variety of ways.

- **Window functions**: Combine aggregate functions with window functions for advanced analysis over partitioned data.

- **Nested GROUP BY**: Explore nested GROUP BY clauses to conduct more in-depth and granular data analysis.

- **Common table expressions (CTEs)**: Make complex queries easier to read and maintain by reducing the number of CTEs in your SQL code.

Looking Ahead

The next chapter, "Ordering the Plot with ORDER BY and LIMIT," explores how to sort and filter query results efficiently. You will be able to organize data in a meaningful way once you have mastered this operation. You can prioritize key information and limit results to focus on the most relevant data points.

Test Your Skills

A database has three tables—gym_memberships (Member_ID, Name, Membership_Type, Join_Date, and City), classes (Class_ID, Class_Name, Category, Class_Date, and Trainer_Name), and attendance (Attendance_ID, Member_ID, Class_ID, Attendance_Date, and Class_Fee).

1. Write an SQL query using window functions to calculate the total class fees and rank for each member based on their class attendance. Class name, member ID, total class fees, and rank must all be shown.

2. Write a query using ROWS to calculate a moving average of class fees over the last three classes attended by each member, including the current one. Provide the member ID, class name, attendance date, and an average of class fees.

3. Write a query using CTEs to find members who have spent more than $500 on classes. The member's name, amount of class fees, and city must be shown.

4. Write a query using a nested GROUP BY to find the top two members who spent the most on classes within each membership type. Show the member ID, membership type, and total class fees.

5. Write a query to calculate the cumulative class fees for each member, ordered by class date, using a window function. Provide a table that shows the member ID, class name, class date, class fee, and cumulative fees.

CHAPTER 6

Ordering the Plot with ORDER BY and LIMIT

Data analysis requires that analysts order the data to provide clarity, insight, and effective communication. ORDER BY clauses in SQL make patterns and trends more visible, helping analysts prioritize findings and detect outliers. When you sort data appropriately, patterns and relationships within the data become clearer, allowing you to make better decisions and take action. For example, consider a list of product sales without any order. The task of identifying top-performing products, tracking trends, or even spotting anomalies would be very challenging. Data sorting provides clarity and enables you to identify patterns and make informed decisions quickly. LIMIT clauses complement this by focusing on the most relevant data points, such as top performers or recent entries, so it is easier to focus on what is most important.

Introduction to ORDER BY

ORDER BY clauses in SQL are used to sort results according to a column or columns. By controlling the order in which the data is displayed, you can identify trends, top performers, or specific patterns more easily. The basic syntax of the ORDER BY clause is as follows:

```
SELECT column1, column2, ...
FROM table_name
ORDER BY column1 [ASC | DESC], column2 [ASC | DESC];
```

© Hamed Tabrizchi 2025
H. Tabrizchi, *Narrative SQL*, https://doi.org/10.1007/979-8-8688-1560-7_6

Here, `column1` and `column2` are the columns you want to sort. You can choose ASC for ascending order (the default) or DESC for descending order. Sorting by multiple columns is also possible. It should be noted that if two rows have the same value in `column1`, SQL will use `column2` as the secondary sorting criteria. This flexible ordering helps present data more effectively. With `ORDER BY`, you can quickly create well-structured outputs, whether you want a list of top-selling items or employees arranged by date of hire.

Ordering Data in Real-World Scenarios

`ORDER BY` clauses are essential to organizing data effectively in SQL queries in various real-world scenarios. Table 6-1 illustrates how to use the `ORDER BY` clause to identify top-selling products, prioritize orders, highlight top-scoring students, or organize customer feedback. These scenarios demonstrate the flexibility of `ORDER BY` for sorting by multiple columns to achieve specific goals, such as ranking items, identifying trends, or prioritizing urgent items. For each scenario, an example SQL query is included, showing you how to construct `ORDER BY` clauses. Table 6-1 also explains how each query is structured and what kind of output it produces.

Table 6-1. *ORDER BY Clauses*

Real-World Scenario	Role of ORDER BY Clause	SQL Query Example	Description
Top-selling products in an e-commerce store	Identify the best-selling products.	`SELECT product_name, total_sales FROM products ORDER BY total_sales DESC;`	Retrieves the top best-selling products by descending `total_sales`.
Employee salaries in a company	Display salaries in descending order.	`SELECT employee_name, salary FROM employees ORDER BY salary DESC;`	Lists all employees, sorted by salary from highest to lowest.

(*continued*)

Table 6-1. (*continued*)

Real-World Scenario	Role of ORDER BY Clause	SQL Query Example	Description

Customer feedback sorted by date	Show the most recent feedback first; useful for addressing current customer concerns or issues.	`SELECT customer_id, feedback, feedback_ date`
		`FROM feedbacks`
		`ORDER BY feedback_ date DESC;`
			Displays all customer feedback, showing the latest entries first.

Order fulfillment priority	Prioritize orders based on urgency (order date or delivery deadline).	`SELECT order_id, customer_name, order_ date`
		`FROM orders`
		`ORDER BY delivery_ date ASC;`
			Lists orders sorted by delivery_date to prioritize the earliest deadlines.

Top scoring students in an exam	Highlight students with the highest scores.	`SELECT student_name, score`
		`FROM exam_results`
		`ORDER BY score DESC;`
			Retrieves the top students based on exam scores in descending order.

Most recent articles on a blog	Display articles based on their publication date, ensuring that the newest content is shown to users first.	`SELECT title, publication_date`
		`FROM articles`
		`ORDER BY publication_ date DESC;`
			Lists all blog articles, starting with the most recently published ones.

Financial transactions sorted by amount	Identify large transactions; useful for fraud detection.	`SELECT transaction_ id, transaction_ amount`
		`FROM transactions`
		`ORDER BY transaction_ amount DESC;`
			Displays all transactions, sorted by amount from highest to lowest.

(*continued*)

Table 6-1. (*continued*)

Real-World Scenario	Role of ORDER BY Clause	SQL Query Example	Description
Customers with the highest lifetime value	Identify valuable customers based on the total revenue generated.	`SELECT customer_id, lifetime_value FROM customers ORDER BY lifetime_value DESC LIMIT 10;`	Retrieves the top ten customers by lifetime value, in descending order.
Movies sorted by rating	Show the highest-rated movies first; useful for recommendation systems or reviews.	`SELECT movie_name, rating FROM movies ORDER BY rating DESC;`	Lists all movies, starting from the highest to the lowest rating.
Latest updates on social media posts	Prioritize the newest posts, comments, or reactions to keep the feed relevant and current.	`SELECT post_id, content, post_date FROM posts ORDER BY post_date DESC;`	Shows all social media posts, starting with the latest ones.
Top trending hashtags	Identify current trends and most-used hashtags in real time.	`SELECT hashtag, usage_count FROM hashtags ORDER BY usage_count DESC;`	Retrieves the top hashtags sorted by the number of times they've been used.

Table 6-1 illustrates how data ordering can provide insights into a variety of realistic scenarios. When records are ordered by date, time-series data can reveal trends over time, such as seasonal peaks. Sales managers can identify growth trends and declining periods by analyzing monthly sales figures ordered chronologically, enabling them to take proactive measures. Sorting financial transactions according to transaction amount helps identify anomalies, such as unusually high transactions. The top performers can be identified by sorting employee performance based on metrics such as project completion time. In this way, analysts can focus on critical areas, gain insights, and make informed decisions based on visible patterns and trends.

Introduction to LIMIT

The LIMIT clause specifies how many rows should be returned by an SQL query. It's especially useful when you only need a subset of data, such as the top results or a sample of entries, without retrieving the entire dataset. LIMIT allows you to control how much data is displayed, so you can focus on the most relevant records. It should be noted that the LIMIT clause is standard in PostgreSQL, MySQL, and SQLite, but other databases may require alternative syntax. Here is a basic explanation of how the LIMIT clause is structured:

```
SELECT column1, column2, ...
FROM table_name
ORDER BY column_name [ASC | DESC]
LIMIT number_of_rows;
```

SELECT column1, column2, ... specifies which columns you want to retrieve from the table, FROM table_name indicates the table from which you are querying data, and ORDER BY column_name [ASC | DESC] specifies the column(s) used to sort the rows before applying the LIMIT. Analysts often combine ORDER BY with LIMIT to ensure the most relevant data is selected, and LIMIT number_of_rows restricts the result set to a specified number of rows. For example, if you use LIMIT 5, only the top five rows will be returned.

Pagination with OFFSET and LIMIT

For user interfaces like web applications, pagination is essential to efficiently display data by dividing large datasets into smaller chunks. Consider browsing through a catalog of thousands of products—it's easier to view 10 or 20 items per page rather than loading the entire list at once. OFFSET and LIMIT clauses in SQL are used to paginate. OFFSET specifies the starting point for the retrieval, while LIMIT specifies how many rows are retrieved. This means that before the LIMIT clause takes effect, users can skip a specified number of rows. Generally, OFFSET is supported in PostgreSQL, MySQL, and SQLite. Other databases like SQL Server and Oracle use different approaches, including TOP, ROWNUM, or OFFSET-FETCH.

> **Note**   When the OFFSET clause is used, the result set is skipped a specified number of rows before returning anything. In addition to pagination, this lets you jump to any position within the dataset, starting at a given index.

The following is a brief explanation of how OFFSET and LIMIT work.

```
SELECT column1, column2, ...
FROM table_name
ORDER BY column_name [ASC | DESC]
LIMIT number_of_rows OFFSET start_position;
```

As mentioned, the LIMIT clause determines the number of rows to display. OFFSET specifies how many rows should be skipped before LIMIT is applied. In most cases, start_position is calculated based on the current page number and number of rows. Table 6-2 illustrates how OFFSET and LIMIT are used in PostgreSQL for pagination.

Table 6-2. *OFFSET and LIMIT for Pagination*

Scenario	Common Use	OFFSET	LIMIT	Query	Description
E-commerce product listing	Paginating products for a user browsing	Varies by page number	Typically 10-50	SELECT * FROM products ORDER BY product_id LIMIT 20 OFFSET (page_ number - 1) * 20;	Here, the expression (page_number - 1) * 20 calculates the starting point (OFFSET) for each page in pagination. By subtracting 1 from page_number, the formula ensures that the first page starts with an offset of 0, while each subsequent page starts 20 records farther.

(*continued*)

Table 6-2. (*continued*)

Scenario	Common Use	OFFSET	LIMIT	Query	Description
Social media feed	Loading posts in a user's feed	Based on scroll count	10-20	`SELECT * FROM posts WHERE user_id = user_id ORDER BY post_date DESC LIMIT 10 OFFSET (scroll_ count - 1) * 10;`	Loads a specific number of recent posts for the user's feed. The query paginates by setting an offset based on the scroll count.
News website	Displaying paginated news articles	Calculated per page	5-15	`SELECT * FROM articles ORDER BY publish_date DESC LIMIT 10 OFFSET (page_ number - 1) * 10;`	Fetches a page of articles ordered by the most recent publish date, adjusting for the specified page_ number.
Admin dashboard logs	Viewing logs or reports	By requested page	Configurable, 20-100	`SELECT * FROM logs ORDER BY timestamp DESC LIMIT 50 OFFSET (page_ number - 1) * 50;`	Loads log entries in descending timestamp order for an admin view, with each page containing 50 records, paginated by page number.
File or data exports	Exporting data with pagination limits	Continuously updated	User-specified or 1,000	`SELECT * FROM data ORDER BY record_id LIMIT 1000 OFFSET (batch_ number - 1) * 1000;`	Exports data in batches of 1,000 records; useful for large datasets. Adjusts the offset based on batch to maintain memory efficiency.

As shown in Table 6-2, each query is designed to handle large datasets efficiently by using LIMIT and OFFSET to control data retrieval. This is done to ensure smooth user experiences across various applications.

Note   Expressions like (page_number - 1) * page_size are commonly used to calculate the starting point, or the *offset*, for paginated data. In general, page_number - 1 adjusts for the fact that pages typically start at 1 (e.g., Page 1, Page 2, etc.), but data offsets start at 0. By subtracting 1, you align the page number with the data index. Also, page_size multiplies the adjusted page number by the number of records desired per page (page_size). This shifts each page forward by exactly one page's worth of records. For example, if page_size is 10:

- Page 1's offset is (1 - 1) * 10 = 0 (starts at the first record).
- Page 2's offset is (2 - 1) * 10 = 10 (starts at the 11th record).

In database queries, this approach is widely used to navigate data in manageable portions.

OFFSET can generally impact performance for large datasets, as it skims rows internally. For large offsets, it is better to consider indexed pagination, which is introduced in Chapter 9.

The First Story: Highway Construction and a Traffic Situation

When a new highway was constructed in a bustling city, a talented data analyst named Pedro was called to investigate the impact on traffic. City council members were eager to determine whether the highway had reduced traffic issues as planned. Pedro uses a dataset containing vehicle counts, travel times, and congestion levels on the main routes. With this data, Pedro planned to identify trends, compare changes, and determine if the highway had brought the relief the city so badly needed. His journey began with defining queries that would reveal hidden patterns and insights to inform council future decisions. These are the questions Pedro seeks answers to:

1. What were the top five most congested routes before and after the highway's construction?

2. How did the average travel time change for routes before and after the highway was built?

3. What routes show significant traffic flow improvements based on the data time provided on the dataset?

4. How has the highway impacted routes that were previously highly congested? Are they still among the top congested routes post-construction?

Pedro uses two tables to compare traffic conditions before and after the highway was constructed. These tables include fields that can help answer the analysis questions provided earlier.

Table 6-3 contains traffic data before the highway was constructed. It includes information on specific routes, vehicle counts, average travel times, and congestion levels, allowing Pedro to assess baseline conditions.

Table 6-3. *The TrafficData_Before Table*

Route_ID	Route_Name	Vehicle_Count	Avg_Travel_Time	Congestion_Level	Data_Date
1	Valley Road	2500	45 min	High	2022-06-01
2	Riverside Ave	1800	38 min	Moderate	2022-06-01
3	Main Street	3000	50 min	High	2022-06-01
4	5th Avenue	2200	40 min	High	2022-06-01
5	Oakwood Blvd	1600	35 min	Moderate	2022-06-01
6	Park Lane	1300	30 min	Low	2022-06-01
7	Maple Drive	1700	32 min	Low	2022-06-01
8	Birch Street	1900	42 min	Moderate	2022-06-01
9	Sunset Blvd	2800	48 min	High	2022-06-01
10	Cedar Road	2100	37 min	Moderate	2022-06-01

Table 6-4 records traffic data for the same routes after the highway's construction. It includes updated vehicle counts, average travel times, and congestion levels. Pedro will use this data to determine the improvement in traffic flow.

Table 6-4. *The TrafficData_After Table*

Route_ID	Route_Name	Vehicle_Count	Avg_Travel_Time	Congestion_Level	Data_Date
1	Valley Road	2000	30 min	Moderate	2023-06-01
2	Riverside Ave	1500	28 min	Low	2023-06-01
3	Main Street	2500	40 min	Moderate	2023-06-01
4	5th Avenue	1800	35 min	Moderate	2023-06-01
5	Oakwood Blvd	1400	25 min	Low	2023-06-01
6	Park Lane	1200	27 min	Low	2023-06-01
7	Maple Drive	1600	29 min	Low	2023-06-01
8	Birch Street	1700	31 min	Moderate	2023-06-01
9	Sunset Blvd	2300	37 min	Moderate	2023-06-01
10	Cedar Road	1900	32 min	Moderate	2023-06-01

In order to get insight into of traffic conditions before and after the construction of the highway, Pedro analyzed this data.

The following query aims to find the top five most congested routes before and after the highway was built. Pedro crafted two SQL queries: one for the `TrafficData_Before` table and another for the `TrafficData_After` table. These tables held data from different time periods, but Pedro ensured that both queries shared the exact same structure, allowing him to directly compare the results:

```
-- Top 5 most congested routes before the highway construction
SELECT Route_Name, Vehicle_Count, Avg_Travel_Time, Congestion_Level
FROM TrafficData_Before
ORDER BY Congestion_Level DESC, Vehicle_Count DESC
LIMIT 5;

-- Top 5 most congested routes after the highway construction
SELECT Route_Name, Vehicle_Count, Avg_Travel_Time, Congestion_Level
FROM TrafficData_After
ORDER BY Congestion_Level DESC, Vehicle_Count DESC
LIMIT 5;
```

Both queries order the data by Congestion_Level, where High is more congested than Moderate or Low. He then ordered the data by Vehicle_Count in descending order to find the most heavily trafficked routes. It is possible to restrict the results to only the top five congested routes by specifying LIMIT 5.

It should be noted that, when using the ORDER BY clause in SQL, certain values have a particular order if they are recognized. Here, due to the alphabetical order of H, M, and L, High will come before Moderate, and Moderate will come before Low. For this reason, the upper query works properly. But where it is necessary to define a specific order, queries must be written differently. The CASE statement can be used to explicitly specify the order. To define congestion levels from High to Moderate to Low, consider the following query:

```
SELECT Route_Name, Vehicle_Count, Avg_Travel_Time, Congestion_Level
FROM TrafficData_After
ORDER BY
  CASE
    WHEN Congestion_Level = 'High' THEN 1
    WHEN Congestion_Level = 'Moderate' THEN 2
    WHEN Congestion_Level = 'Low' THEN 3
  END,
  Vehicle_Count DESC
LIMIT 5;
```

As a result, each congestion level has a numerical value. High has the highest priority (1), followed by Moderate (2), and then Low (3). This method ensures that the desired order is achieved regardless of the sorting method.

Note   In most comparative analyses, using the same query structure on different datasets is a common and effective method because it allows direct comparison of results across different time periods or conditions. By maintaining a similar query structure, analysts ensure that data variations are due to data itself, rather than logical differences in the query. As a result of this method, findings are more reliable and valid, since this method reduces the potential for unintentional bias.

Table 6-5 shows the top five most congested routes before the highway was built, with Main Street having the highest traffic volume and congestion.

Table 6-5. *Congested Routes Before Construction*

Route_Name	Vehicle_Count	Avg_Travel_Time	Congestion_Level
Main Street	3000	50 min	High
Valley Road	2500	45 min	High
Sunset Blvd	2800	48 min	High
5th Avenue	2200	40 min	High
Riverside Ave	1800	38 min	Moderate

Table 6-6 shows the top five most congested routes after the highway construction, indicating a reduction in congestion and travel times.

Table 6-6. *Congested Routes After Construction*

Route_Name	Vehicle_Count	Avg_Travel_Time	Congestion_Level
Main Street	2500	40 min	Moderate
Sunset Blvd	2300	37 min	Moderate
Valley Road	2000	30 min	Moderate
5th Avenue	1800	35 min	Moderate
Cedar Road	1900	32 min	Moderate

The following query finds the average travel time change before and after the highway was built. By performing these queries, Pedro can calculate the average travel time for all paths before and after the highway was constructed, which will help him determine if there has been a noticeable improvement in the travel time throughout the city.

```
-- Average travel time before construction
SELECT AVG(CAST(SUBSTRING(Avg_Travel_Time FROM '^[0-9]+') AS INT)) AS Avg_
Travel_Time_Before FROM TrafficData_Before;
```

```
-- Average travel time after construction
SELECT AVG(CAST(SUBSTRING(Avg_Travel_Time FROM '^[0-9]+') AS INT)) AS Avg_
Travel_Time_After FROM TrafficData_After;
```

The queries use *casting* to convert `Avg_Travel_Time` values from strings style like `'30 min'` to integers. This conversion is essential because the AVG function requires numerical input to compute the average. By extracting the numeric portion of the travel time and converting it to an integer, Pedro can effectively analyze the data and draw meaningful conclusions about the impact of the highway on travel times.

Tables 6-7 and 6-8 show the average travel times across all routes before and after the highway construction, showing a significant reduction in travel time post-construction.

Table 6-7. *The Average Travel Times Across All Routes (Before)*

Avg_Travel_Time_Before
39.7

160

Table 6-8. *The Average Travel Times Across All Routes (After)*

Avg_Travel_Time_After
31.4

This query is provided to find traffic flow improvements based on the data time provided in the dataset:

```
SELECT Route_Name, Vehicle_Count, Avg_Travel_Time, Congestion_Level
FROM TrafficData_After
WHERE Data_Date BETWEEN '2023-05-01' AND '2023-06-01'
ORDER BY Avg_Travel_Time ASC
LIMIT 5;
```

This query filters data from the specific range of '2023-05-01' to '2023-06-01' and sorts by Avg_Travel_Time in ascending order to find routes with shorter travel times, indicating improved traffic flow. LIMIT 5 restricts the output to the top five routes with the best improvements.

Table 6-9 shows routes with the most improved travel times over the mentioned 30 days, with routes like Oakwood Blvd and Park Lane experiencing the shortest travel times and lowest congestion levels.

Table 6-9. *Routes with the Most Improved Travel Times*

Route_Name	Vehicle_Count	Avg_Travel_Time	Congestion_Level
Oakwood Blvd	1400	25 min	Low
Park Lane	1200	27 min	Low
Riverside Ave	1500	28 min	Low
Maple Drive	1600	29 min	Low
Valley Road	2000	30 min	Moderate

To find the impacted routes that were previously highly congested and see if they are still among the top congested routes post-construction, Pedro used the following query:

```
SELECT Route_Name, Congestion_Level
FROM TrafficData_Before
WHERE Congestion_Level = 'High'
UNION
SELECT Route_Name, Congestion_Level
FROM TrafficData_After
WHERE Congestion_Level = 'High';
```

This query combines routes with a High congestion level from before and after the highway construction. The UNION operator displays routes that were previously very busy and indicates if they remain busy.

Note   The UNION operator is used to combine the results of two or more SELECT queries into a single result set. Each SELECT statement within a UNION must have the same number of columns in the same order and with compatible data types. When using UNION, only unique rows are returned, so any duplicate rows between the queries will be removed automatically. To retain duplicates, you can use UNION ALL. Other set operators in PostgreSQL, including UNION, UNION ALL, INTERSECT, INTERSECT ALL, EXCEPT, and EXCEPT ALL, are discussed in the next chapters.

Table 6-10 shows routes with a high congestion level before the highway was built and indicates if any remain highly congested after construction; none remain High post-construction, indicating an improvement.

Table 6-10. *Remain Highly Congested Routes After Construction*

Route_Name	Congestion_Level
Main Street	High
Valley Road	High
Sunset Blvd	High

Pedro used SQL to analyze traffic data before and after highway construction. Through careful sorting, filtering, and pagination of large datasets, Pedro produced insights that demonstrated an improvement in overall traffic conditions. His analysis provided city officials with clear evidence that the highway had achieved its goal of alleviating congestion. This validated the city's investment and set a data-driven standard for future infrastructure projects.

Customizing Your Sorting: Advanced Use Cases of ORDER BY

Case Sensitivity and Sorting Strings

In SQL, case sensitivity affects the order of results, especially when comparing uppercase and lowercase characters. If uppercase letters are not handled properly, SQL databases may sort uppercase letters lowercase, which may lead to unexpected results. Using the COLLATE clause, analysts can perform case-insensitive or case-sensitive sorting. Most databases allow defining how text should be compared and sorted.

What Is COLLATE?

In SQL, you can apply rules to compare strings and sort them by using the COLLATE clause, such as case sensitivity (for example, distinguishing A from a), accent sensitivity (for multilingual datasets), and language or regional conventions for sorting.

Note   A *collation* in SQL is a set of rules that determines how text strings are compared, ordered, and sorted. Collations are particularly important when dealing with languages, as different languages have unique rules for alphabetical ordering and character comparison. For instance, a collation controls:

1. **Case sensitivity:** Determines if uppercase and lowercase letters are treated as equal (case-insensitive) or different (case-sensitive).

2. **Accent sensitivity:** Specifies if accents or diacritical marks on characters (like é vs. e) are considered unique.

3. **Locale-specific sorting rules:** Defines sorting orders based on language or regional rules, which might differ. For instance, in some languages, special characters like ø might be sorted differently.

Collation in PostgreSQL

In PostgreSQL, collations can be applied at various levels, including:

- **Database level**: Defines the default collation for all string columns within a database.

- **Column level**: Sets a specific collation for a particular column in a table, overriding the database's default.

- **Query level**: Allows using COLLATE in a query to temporarily change the sorting behavior for that specific operation.

PostgreSQL provides many predefined collations based on locale settings, allowing developers to select collations suited for particular languages and case-sensitivity needs.

Using COLLATE

In PostgreSQL, COLLATE is a clause allowing you to specify collation order for a query or operation. This determines how text strings are sorted and compared. Collation affects character ordering and case sensitivity. COLLATE is used for sorting and comparing strings, especially suited for locale-specific ordering. For example, in German, ä and a might be sorted differently than in English. COLLATE can be used with CREATE TABLE, SELECT, or any text-manipulation operation. PostgreSQL comes with several built-in collations. You can list all available collations by using the following options.

COLLATE with CREATE TABLE

```
CREATE TABLE employees (
    name TEXT COLLATE "en_US",
    department TEXT COLLATE "de_DE"
);
```

Here, the name column will be sorted using U.S. English rules, and the department column will be sorted using German rules.

COLLATE in SELECT Query

```
SELECT * FROM employees ORDER BY name COLLATE "fr_FR";
```

This query orders the results by name using French collation rules, regardless of the default collation.

Example: Case Sensitivity with Different Collations

As a first step, you'll create a users table with a column for usernames. You will compare and order usernames with different case types to see how they work. The following data is stored in the users table, with a username column:

```
CREATE TABLE users (
    username TEXT
);

INSERT INTO users (username) VALUES
('Alice'),
('alice'),
('Bob'),
('bob');
```

This table now has mixed-case entries: Alice, alice, Bob, bob, Charlie, and charlie. Using COLLATE, you can compare results when you query the table with a case-sensitive collation versus a case-insensitive collation.

Case-Sensitive Ordering

To see how case-sensitive ordering works, you'll use the C collation, which is usually case-sensitive and orders uppercase letters before lowercase letters.

```
SELECT * FROM users
ORDER BY username COLLATE "C";
```

Table 6-11 shows the results of this query.

Table 6-11. *Case-Sensitive Ordering Output*

Username
Alice
Bob
Charlie
Alice
Bob
Charlie

In this case, uppercase names (Alice, Bob, Charlie) are listed before lowercase names (alice, bob, charlie). This ordering is due to C collation treating uppercase letters as distinct from lowercase letters and sorting them first.

Case-Insensitive Ordering

The case-insensitive collation en_US.utf8 treats uppercase and lowercase characters equally for sorting.

```
SELECT * FROM users
ORDER BY username
COLLATE "en_US.utf8";
```

In this case, as shown in Table 6-12, Alice and alice are grouped together, as are Bob with bob, and Charlie with charlie. This shows that the en_US.utf8 collation ignores case differences when ordering, resulting in more "human-friendly" sorting.

Table 6-12. *Case-Insensitive Ordering Output*

username
Alice
alice
Bob
bob
Charlie
charlie

Sorting NULL Values

In PostgreSQL, handling NULL values during sorting can be crucial, especially when organizing data that might have missing entries. By default, NULL values are considered lower than any other value in ascending order and higher in descending order. However, PostgreSQL provides explicit options for sorting NULL values either first or last, allowing you to control their placement within ordered results.

Strategies for Ordering NULL Values: NULLS FIRST and NULLS LAST

There are two strategies for sorting NULL values. The first one is NULLS FIRST and the second one is NULLS LAST.

NULLS FIRST: Place all NULL values at the beginning of the result set. This is often useful when you want missing values to appear prominently, for example, to prioritize records that need attention.

```
ORDER BY column_name NULLS FIRST;
```

NULLS LAST: Place all NULL values at the end of the result set. This is useful when NULL values are less relevant, and you want actual data values to appear first.

```
ORDER BY column_name NULLS LAST;
```

Example: Customer Data with NULLs in the Purchase History

Take the example of a table customer containing customer information, shown in Table 6-13, including a last_purchase_date column. In some cases, last_purchase_date may be NULL, indicating that the customer hasn't purchased anything yet.

Table 6-13. *The Customers Table*

customer_id	customer_name	last_purchase_date
1	Alice	2023-06-15
2	Bob	
3	Charlie	2023-07-20
4	Diana	
5	Eve	2023-05-10

Sorting with NULLS FIRST

To list customers with the most recent purchases first and those who haven't made a purchase appearing at the top, use this query:

```
SELECT * FROM customers
ORDER BY last_purchase_date
DESC NULLS FIRST;
```

Table 6-14 shows that NULLs are sorted first in the sorting process.

Table 6-14. *Customer Sorting with NULLS FIRST*

customer_id	customer_name	last_purchase_date
2	Bob	NULL
4	Diana	NULL
3	Charlie	2023-07-20
1	Alice	2023-06-15
5	Eve	2023-05-10

In this case, NULL values appear at the top due to NULLS FIRST, followed by recent purchase dates in descending order.

Example 2: Sorting with NULLS LAST

You can use NULLS LAST to display customers who haven't bought anything yet:

```
SELECT * FROM customers
ORDER BY last_purchase_date
DESC NULLS LAST;
```

Table 6-15 shows that the NULLs are sorted last in the sorting process.

Table 6-15. *Customer Sorting with* NULLS LAST

customer_id	customer_name	last_purchase_date
3	Charlie	2023-07-20
1	Alice	2023-06-15
5	Eve	2023-05-10
2	Bob	NULL
4	Diana	NULL

First, non-NULL dates appear in descending order, with NULL values at the bottom because of NULLs LAST.

The purpose of NULLS FIRST is to prioritize NULL values at the top, whereas NULLS LAST moves NULL values to the bottom of the ordered results. You can control how NULL values appear in your results, which makes the data presentation more meaningful based on context.

Common Pitfalls and Best Practices

Avoiding Ambiguous Ordering: Always Clarify Column Names

When sorting data, ambiguity in column names can lead to unexpected results or even query errors. This commonly occurs in queries involving multiple tables (like joins) or complex subqueries, where columns might share similar names.

In PostgreSQL, when you query multiple tables with overlapping column names, like `customer_id` in both customers and orders, the database engine cannot immediately determine which table's column you're referring to if the column name is used without an alias or table name prefix. This is especially problematic in sorting (`ORDER BY`) and filtering (`WHERE`) clauses. There are three common reasons why ambiguity occurs.

1. **Column name overlap**: When multiple tables are joined, they may have columns with the same name, such as `customer_id`. Without specifying a table name or alias, PostgreSQL can't determine which column you mean.

2. **ORDER BY clause requirements**: PostgreSQL expects unambiguous references for sorting because it relies on table column values. If it encounters a column name that exists in multiple tables without an explicit table reference, it doesn't know which one to sort by, causing ambiguity.

3. **Lack of context for aggregation and sorting**: In queries where columns aren't explicitly associated with a table, PostgreSQL lacks the context to identify which data source to pull from, leading to errors or unexpected behavior.

Avoiding Ambiguity

Whenever you query or sort on columns that are part of multiple tables, use table names or aliases to identify which table each column belongs to. As a result, PostgreSQL uses the correct data and eliminates ambiguity.

An Example of Ambiguity in Column Names When Sorting Data

The Customers table stores information about each customer, including a unique customer_id and name. The Orders table records purchase transactions, including order_id, the customer_id linking orders to specific customers, and the order_date of each transaction. Tables 6-16 and 6-17 enable tracking customer purchases over time, associating each order with a particular customer.

Table 6-16. *The Customers Table*

customer_id	name
1	Alice
2	Bob
3	Charlie
4	Diana
5	Eve

Table 6-17. *The Orders Table*

order_id	customer_id	order_date
1	1	2023-06-15
2	1	2023-07-20
3	2	2023-05-10
4	3	2023-08-01
5	3	2023-07-22
6	4	2023-04-18
7	5	2023-05-25

The purpose of this query is to retrieve `customer_id` and `order_date`, sorted by `order_date`. If the table names or aliases are not used explicitly in the query, PostgreSQL can become confused due to ambiguous column names, particularly when the same column name exists in both tables:

```
SELECT customer_id, order_date
FROM customers
JOIN orders ON customers.customer_id = orders.customer_id
ORDER BY order_date DESC;
```

The `order_date` column exists only in the `Orders` table, but the `customer_id` column is present in both `Customers` and `Orders`. Without explicit clarification, PostgreSQL may not reliably determine the intended source of `order_date` or even confuse `customer_id`. As a result, this error is thrown when this query is executed.

```
psql:commands.sql:36: ERROR:  column reference "customer_id" is ambiguous
LINE 1: SELECT customer_id, order_date
```

By specifying the table name or alias, you remove any potential for ambiguity. Here's the revised query:

```
SELECT customers.customer_id, orders.order_date
FROM customers
JOIN orders ON customers.customer_id = orders.customer_id
ORDER BY orders.order_date DESC;
```

This version explicitly references the `orders.order_date` column in the `ORDER BY` clause, ensuring PostgreSQL knows which table to use for sorting.

This result displays the `order_date` column from the `Orders` table in descending order, as shown in Table 6-18.

Table 6-18. *Descending Order Given the order_date*

customer_id	order_date
3	2023-08-01
3	2023-07-22
1	2023-07-20
1	2023-06-15
5	2023-05-25
2	2023-05-10
4	2023-04-18

Plot Efficiency with ORDER BY and LIMIT

ORDER BY and LIMIT are particularly useful when plotting because they allow you to:

- **Focus on relevant data**: The plot should focus on key insights rather than all data points when analyzing large datasets.

- **Improve plot clarity**: The plot becomes easier to understand by limiting the data to a manageable subset, such as the most relevant or impactful rows.

- **Enhance performance**: Large datasets with thousands of rows can slow down plotting processes and overwhelm visual tools, resulting in longer load times and reduced responsiveness. Using LIMIT with ORDER BY avoids overloading the plot by fetching only a select number of rows. This makes the visualization process faster and more efficient.

- **Highlight trends and outliers**: When plotting specific data points, ordered and limited data makes identifying trends, outliers, and comparisons easier.

In this chapter, a *plot* generally refers to arranging or presenting data in a structured, meaningful way, often as preparation for visualization. Using `ORDER BY` and `LIMIT` clauses helps achieve this by sorting and narrowing down data into the most relevant entries. Despite the power of data storage and querying, third-party visualization tools are available, which make it possible to create visualizations after you have prepared and ordered the data. `ORDER BY` organizes the result set based on specified columns, either in ascending (`ASC`) or descending (`DESC`) order, allowing you to prioritize data. Meanwhile, `LIMIT` constrains the number of rows returned, making it easier to focus on a subset, such as the top results. Together, these queries refine data for clearer, more insightful analysis and presentation.

Summary

This chapter explained the importance of using the `ORDER BY` and `LIMIT` clauses in SQL queries to organize and manage datasets effectively. It highlighted how these clauses help narrow down query results by sorting data in a specific order and limiting the number of rows returned. By limiting and selecting data selectively, analysts can enhance the clarity and performance of data analysis and visualization tasks. These SQL features are demonstrated in the chapter in terms of filtering data for targeted insights, improving visual plot effectiveness, and optimizing query performance.

Key Points

- `ORDER BY` and `LIMIT` help focus on the most relevant entries within large datasets. This allows analysts to prioritize data that directly impacts insights, such as the highest or lowest values in specific columns.

- By limiting the number of rows returned, `ORDER BY` and `LIMIT` make plots more readable, presenting only the most relevant data and reducing visual complexity in graphs or tables.

- `ORDER BY` and `LIMIT` help stakeholders make intelligent decisions without sifting through unnecessary details by returning only the most critical data points.

- Using OFFSET and LIMIT helps maintain an orderly structure for viewing large datasets, as each page displays a consistent number of entries, improving data accessibility and clarity. By reducing the computational load on databases, LIMIT also speeds up query performance and reduces resource consumption.

Key Takeaways

- ORDER BY: Use ORDER BY to sort data by specific columns, prioritizing key insights like the highest or lowest values. This makes it easier to identify trends and patterns in large datasets.

- LIMIT: By applying LIMIT, analysts can restrict the number of rows returned, so that the most relevant data points are returned. This is especially useful in visualizations where only the top or bottom records are needed, reducing clutter and enhancing clarity.

- OFFSET: Combining OFFSET with LIMIT for pagination allows you to retrieve data in manageable chunks and navigate large datasets page-by-page, rather than all at once. By using OFFSET and LIMIT together, pagination becomes possible. This technique supports a more user-friendly presentation of data and ensures efficient performance by loading data incrementally rather than processing it all simultaneously.

Looking Ahead

The next chapter, "Dynamic Dialogues with Subqueries," explores the art of writing subqueries to add depth and dimension to data analysis. By employing nested queries, you can achieve meaningful insights and tackle complex analytical challenges.

Test Your Skills

Jack is a data analyst for the Silverstone City Library, which has an extensive book collection and a large patronage. Library administration is interested in improving the user experience by understanding borrowing patterns, popular titles, and member activity. Using SQL techniques like ORDER BY, LIMIT, and OFFSET, help him to make decisions based on library data. The library database contains a table called BookLoans with the columns shown in Table 6-19.

Table 6-19. *The BookLoans Table*

Column	Description
loan_id	Unique identifier for each book loan
member_id	Unique identifier for each member
book_title	Title of the borrowed book
borrow_date	Date when the book was borrowed
return_date	Date when the book was returned
loan_duration	Duration in days of the loan (calculated as return_date - borrow_date)
borrow_count	Number of times the book has been borrowed

1. Identify the top five most frequently borrowed books in the library collection. Order the results by borrow_count in descending order so that the most borrowed books appear at the top.

2. To help staff monitor recent activity, retrieve a list of books borrowed in the past month, ordered by borrow_date from the most recent to the least. Show only five results per page and use OFFSET to navigate between pages.

3. Determine the top five longest loan durations by members. Order results by loan_duration in descending order to show the longest durations first.

4. Find the books with the fewest borrows and display them on the second page (rows 6-10). Order by borrow_count in ascending order and use OFFSET to skip the first five records.

5. Identify the ten books that were borrowed the most in the last year. Order by borrow_count in descending order and filter results by borrow_date within the past year.

CHAPTER 7

Dynamic Dialogues with Subqueries

This chapter explores the concept of dynamic dialogues in SQL, specifically focusing on subqueries in SQL. As mentioned earlier, subqueries are powerful tools that allow you to nest queries inside one another to perform complex queries. The purpose of this chapter is to provide an overview of subqueries, their types, and narrative examples of their use in dynamic dialogues.

Introduction to Subqueries

In SQL, a *subquery* is a query nested inside another query. There are many ways to use subqueries, including within the SELECT, FROM, WHERE, and HAVING clauses. Depending on their structure, subqueries can return single values, multiple values, or even entire tables. Thus, subqueries can be divided into three general types: *single-row subqueries,* which return a single row and are suitable for use with comparison operators; *multiple-row subqueries*, which provide multiple rows and work with operators such as IN, ANY, and ALL; and *correlated subqueries,* which execute once for each row in the outer query based on column references in the subquery.

The First Story: A Bustling Office

In a busy office, Sarah, a junior data analyst, is assigned the task of investigating the employee and department databases. Sarah's manager needed answers to complex questions about employee salaries, departments, and performance. Sarah wants to write nested SQL subqueries to meet these requirements.

© Hamed Tabrizchi 2025
H. Tabrizchi, *Narrative SQL*, https://doi.org/10.1007/979-8-8688-1560-7_7

Sarah uses Tables 7-1 and 7-2 to extract insights related to her assigned task.

Table 7-1. *The Employees Table*

ID	Name	Salary	department_id
1	John	70,000	1
2	Alice	80,000	2
3	Bob	60,000	1
4	Emma	90,000	3
5	Michael	55,000	2

Table 7-2. *The Departments Table*

ID	Name
1	IT
2	Sales
3	Marketing

The first question is who has the highest salary. A person who is the highest-paid employee at the company is what Sarah is seeking. To find the answer, she used the following single-row subquery:

```
SELECT name
FROM employees
WHERE salary = (SELECT MAX(salary) FROM employees);
```

In this query, the subquery `(SELECT MAX(salary) FROM employees)` calculates the highest salary in the table, and the outer query retrieves the name of the employee with that salary. A single row, shown in Table 7-3, is returned by this query, as you might expect.

Table 7-3. *The Highest Salary*

Name
Emma

The second question is who works in the Sales department. The list Sarah is seeking contains all employees in the Sales department. Sarah used a multiple-row subquery to match department IDs to find the answer:

```
SELECT name
FROM employees
WHERE department_id IN (SELECT id FROM departments WHERE name = 'Sales');
```

The subquery (`SELECT id FROM departments WHERE name = 'Sales'`) retrieves the ID of the Sales department, and the outer query matches this with `department_id` in the Employees table. Table 7-4 shows the results.

Table 7-4. *Who Works in the Sales Department*

Name
Alice
Michael

The third question is who earns above the average salary in their department. The list Sarah is seeking contains all employees who earn more than the average salary for their department. This challenge was addressed by Sarah using a correlated subquery:

```
SELECT emp.name
FROM employees AS emp
WHERE emp.salary > (
    SELECT AVG(sub_emp.salary)
    FROM employees AS sub_emp
    WHERE sub_emp.department_id = emp.department_id
);
```

The subquery (SELECT AVG(sub_emp.salary) FROM employees AS sub_emp WHERE sub_emp.department_id = emp.department_id)) calculates the average salary for each department, dynamically adjusted based on the department_id of each employee in the outer query. The emp alias represents the Employees table in the outer query, and the sub_emp alias represents the same table in the subquery. There is a clear indication of their respective roles in the query by these aliases. Table 7-5 shows which employees earn above the average salary in their department.

Table 7-5. *Employees Earning Above the Average Salary in Their Department*

Name
Alice
John

Dynamic Dialogues with Subqueries

In SQL, dynamic dialogues are used to create interactive and responsive queries that can be modified based on the inputs of the user as well as other conditions. As a result of this dynamic behavior, subqueries play an important role. SQL queries can be created using dynamic dialogues that adapt based on parameters or conditions. As a result, the same query can be executed differently depending on the user's input or the database's state. Adaptability is essential for building robust applications that manipulate and retrieve data in real time.

The Role of Subqueries in Dynamic Dialogues

The term subqueries or nested queries refers to queries embedded within another SQL query. They provide powerful mechanisms for conditional logic and adaptability to dynamically filter, aggregate, or transform data. A subquery integrated into a dynamic dialogue allows conditional logic, real-time adaptation, and enhanced modularity. Table 7-6 provides a brief description of conditional logic, real-time adaptation, and enhanced modularity in subqueries.

Table 7-6. *Conditional Logic, Real-Time Adaptation, and Enhanced Modularity in subqueries*

Aspect	Description	Example

Conditional logic	Adjust results dynamically based on thresholds or specific scenarios, such as user preferences.	Retrieve records where a subquery calculates a threshold:
		`SELECT name`
		`FROM employees`
		`WHERE salary > (SELECT AVG(salary) FROM`
		`employees);`

Real-time adaptation	Use parameters or application inputs with subqueries to provide context-aware results.	`SELECT name`
		`FROM employees`
		`WHERE salary > (SELECT MIN(salary) FROM`
		`employees WHERE department_id = 2);`

Enhanced modularity	Break complex operations into smaller, reusable subqueries for clarity and maintainability.	`WITH DeptAvg AS (SELECT department_id,`
		`AVG(salary) AS avg_salary FROM employees`
		`GROUP BY department_id) SELECT e.name`
		`FROM employees e JOIN DeptAvg d ON`
		`e.department_id = d.department_id WHERE`
		`e.salary > d.avg_salary;`

As shown in Table 7-6, conditional logic allows dynamic changes to query results based on thresholds, for example, comparing employee pay to the average. Real-time adaptation, on the other hand, incorporates user input and parameters, ensuring outcomes that are customized to the situation. For instance, you can apply a minimum wage filter to workers in a particular department. In addition, it is possible to use a common table expression (CTE) to divide large queries into parts that can be reused separately; this method simplifies large queries. A departmental average salary can be calculated, for example. In this section, these subqueries are illustrated with examples to demonstrate their flexibility, efficiency, and maintainability.

Introduction to Subqueries as Conversational Elements

As mentioned, subqueries are SQL queries embedded inside another query. As such, they can be referred to as "conversational elements" since they permit queries to interact in a dynamic manner, thus enabling a flow of information between queries. Follow-up questions are often asked in a conversation to obtain more specific information. Subqueries allow SQL queries to build upon each other, enabling more complex data retrieval by refining or modifying the main query based on the results of another query.

Subqueries in SQL allow queries to interact and share information dynamically. There are various types of subqueries, each playing a different role in enhancing data retrieval flow. Single-row subqueries provide direct, one-time answers to questions, while multi-row subqueries return a list of possible responses to broader queries. A multi-column subquery returns multiple columns and rows, typically when comparing tuples, pairs, and groups. Correlated subqueries engage in a continuous back-and-forth conversation with the outer query, as they depend on each row's context for execution. Uncorrelated subqueries are executed only once, and their results are used by the outer query. Lastly, the subqueries in the FROM clause act as a foundational context, preparing summarized data that the main query can further analyze.

By breaking down tasks into smaller, more manageable steps, SQL can build complex, refined queries. The following sections include a more detailed explanation of the types of subqueries, using query examples based on the data provided in Tables 7-7 and 7-8.

Table 7-7. *The Employees Table*

employee_id	name	department_id	job_id	Salary
101	Alice Johnson	1	J001	55000
102	Bob Smith	2	J002	48000
103	Charlie Davis	1	J003	60000
104	David Brown	3	J004	72000
105	Emma Wilson	2	J001	45000
106	Fiona Clark	4	J005	80000
107	George Miller	1	J003	61000
108	Hannah White	4	J006	85000
109	Ian Thompson	3	J002	70000
110	Julia Lewis	2	J004	52000

Table 7-8. *The Departments Table*

department_id	department_name	Location
1	Sales	New York
2	Marketing	Chicago
3	IT	San Francisco
4	HR	Boston
5	Operations	Los Angeles
6	Finance	New York
7	Logistics	Chicago
8	R&D	San Francisco
9	Legal	Boston
10	Customer Support	Los Angeles

Single-Row Subqueries

In a single-row subquery, only one row and one column of data are returned. It is often used when the outer query expects a single value to compare against, such as when using comparison operators like =, <, >, or !=. If the subquery returns more than one row, an error will occur. For instance, the following nested query considers a single-row subquery to find the names of all employees in the Sales department.

```
SELECT name
FROM employees
WHERE department_id = (SELECT department_id FROM departments WHERE
department_name = 'Sales');
```

The inner query finds the `department_id` of the `'Sales'` department, and the outer query retrieves the names of employees in that department.

As shown in Tables 7-9 and 7-10, the inner query returns the `department_id` for the Sales department, and the outer query retrieves the names of employees with `department_id = 1`.

Table 7-9. *The Output of Inner Query*

department_id
1

Table 7-10. *The Output of the Outer Query*

name
Alice Johnson
Charlie Davis
George Miller

Multi-Row Subqueries

A multi-row subquery returns multiple rows but usually a single column. These subqueries are used with operators like IN, ANY, or ALL to match one or more values from the subquery's results. To find the names of all employees who work in any department located in New York, the following nested query considers a multi-row subquery.

```
SELECT name
FROM employees
WHERE department_id IN (SELECT department_id FROM departments WHERE
location = 'New York');
```

The subquery retrieves the IDs of all departments located in 'New York', and the outer query finds employees working in those departments.

Tables 7-11 and 7-12 demonstrate, in the inner query, the IDs of the New York departments are returned. In the outer query, the names of employees in departments with IDs 1 or 6 are returned.

Table 7-11. *The Output of Inner Query*

department_id
1
6

Table 7-12. *The Output of Outer Query*

name
Alice Johnson
Charlie Davis
George Miller

Multi-Column Subqueries

A multi-column subquery returns multiple columns and rows. It is typically used with composite comparisons involving tuples, pairs, or groups of values, often combined with operators like IN or in JOIN conditions. For instance, the following nested query aims to find the names of employees who have both the same department_id and job_id as current open positions in the company.

```
SELECT name
FROM employees
WHERE (department_id, job_id) IN (SELECT department_id, job_id FROM job_
openings WHERE status = 'Open');
```

The subquery retrieves pairs of department_id and job_id where jobs are open, and the outer query finds employees who match those pairs.

Table 7-13 shows the data in the table called job_openings.

Table 7-13. *The job_openings Table*

department_id	job_id	Status
1	J003	Open
4	J006	Open

Subqueries retrieve open jobs, and the outer queries match employees with department-job pairs. See Tables 7-14 and 7-15.

Table 7-14. *The Inner Query Output*

department_id	job_id
1	J003
4	J006

Table 7-15. *The Outer Query Output*

name
Charlie Davis
George Miller
Hannah White

Correlated Subqueries

A correlated subquery depends on the outer query for its execution. It is evaluated once for each row processed by the outer query. Correlated subqueries are more like back-and-forth conversations, where the inner query continuously references columns from the outer query. As an example, the following query compares the salary of each employee with the average salary within their own department, not the average salary throughout the entire organization, in order to identify employees who earn more than the average salary within their own department.

```
SELECT e.name
FROM employees e
WHERE e.salary > (SELECT AVG(salary) FROM employees WHERE department_id =
e.department_id);
```

For each employee in the outer query, the inner query calculates the average salary for that employee's department and compares it to the employee's salary. To assist you in better understanding, Table 7-16 provides an illustration of the average salary for each employee's department.

In the inner query, the average salary for each employee's department is calculated, and in the outer query, as shown Table 7-17, the employee's salary is compared to that average.

Table 7-16. *Illustration of Average Salary for Each Employee's Department*

Name	department_id	salary	avg_department_salary	salary_comparison
Alice Johnson	1	55000	58667	Below Average
Charlie Davis	1	60000	58667	Above Average
George Miller	1	61000	58667	Above Average
Emma Wilson	2	45000	48333	Below Average
Bob Smith	2	48000	48333	Below Average
Julia Lewis	2	52000	48333	Above Average

Table 7-17. *Output of the Correlated Subquery*

Name
Charlie Davis
George Miller
Julia Lewis

Note   This table has been adjusted for better understanding. To write a query to access this table in PostgreSQL, you can write the following query:

```
SELECT
    e.name,
    e.department_id,
    e.salary,
    dept_avg.avg_department_salary,
    CASE
        WHEN e.salary > dept_avg.avg_department_salary THEN 'Above
        Average'
```

```
        ELSE 'Below Average'
    END AS salary_comparison
FROM employees e
JOIN (
    SELECT department_id, AVG(salary) AS avg_department_salary
    FROM employees
    GROUP BY department_id
) dept_avg ON e.department_id = dept_avg.department_id;
```

Uncorrelated Subqueries

An uncorrelated subquery is independent of the outer query. It is executed only once, and its result is used by the outer query. These are more straightforward and they do not reference columns from the outer query. The following query, for instance, first calculates the overall company-wide average salary before showing employees who earn more than it.

```
SELECT name
FROM employees
WHERE salary > (SELECT AVG(salary) FROM employees);
```

The subquery calculates the average salary across all employees once, and the outer query finds employees with a salary above that average.

As Tables 7-18 and 7-19 demonstrate, the inner query calculates average salary across all employees, and the outer query retrieves employees with salaries above $62,800.

Table 7-18. *Average Salary Across All Employees*

avg_salary
62800

Table 7-19. *Employees with Salaries Above $62,800*

name
David Brown
Fiona Clark
Hannah White
Ian Thompson

Subqueries in the FROM Clause

A subquery in a FROM clause is often called a *derived table*. It acts as a temporary table that the outer query can use to retrieve more meaningful insights. This type of subquery is useful for pre-aggregating data before using it in the main query. For example, the following nested query calculates the average salary for each department and provides their names along with their average salaries.

```
SELECT department_name, avg_salary
FROM (SELECT department_id, AVG(salary) AS avg_salary
      FROM employees
      GROUP BY department_id) AS dept_avg
JOIN departments d ON dept_avg.department_id = d.department_id;
```

The subquery calculates the average salary for each department, and the outer query retrieves department names and their corresponding average salaries.

In the inner query, the average salary per department is calculated. A JOIN operation is then performed to join the Departments table. See Tables 7-20 and 7-21.

Table 7-20. *The Average Salary per Department*

department_id	avg_salary
1	58667
2	48333
3	71000
4	82500

Table 7-21. *Department Names and Their*
Corresponding Average Salaries

department_name	avg_salary
Sales	58667
Marketing	48333
IT	71000
HR	82500

Complex Conversations: Nested and Multi-Level Subqueries

In SQL, nested subqueries and multi-level subqueries allow you to solve complex data-retrieval problems from multiple queries by writing multiple layers of queries. Each layer refines the data further, making it more targeted and relevant. This can be seen as similar to a dialogue that involves asking follow-up questions to previous answers, where each layer is dependent on the results of the previous layer.

As mentioned, a nested subquery is a subquery placed inside another subquery or query. In more complex scenarios, multiple subqueries may be nested within one another, resulting in a multi-level subqueries.

Note   A subquery can be nested in a SELECT, FROM, or WHERE clause. Multi-level subqueries require SQL to evaluate the innermost query first and then pass the result to the next query. The outermost query uses the final result to produce the desired output.

General Syntax of Two-Level Subqueries

A two-level subquery is a query where the outer query depends on the result of a subquery, and that subquery itself has another inner subquery.

```
SELECT column1
FROM table1
WHERE column2 = (
    SELECT column3
    FROM table2
    WHERE column4 = (
        SELECT column5
        FROM table3
        WHERE condition
    )
);
```

First, the innermost subquery runs and returns a result, then the middle subquery filters its own data, and finally the outer query filters the main table using the final result.

Complex Multi-Level Subqueries

To write complex nested SQL queries, it's essential to work from the inside out, progressively building and testing each layer to ensure correctness. It is recommended to begin by writing and testing the innermost query, then gradually adding outer layers once you have confirmed the inner portions work properly. Using meaningful table aliases enhances code readability and makes complex queries easier to understand and maintain. The best recommendation is to use descriptive aliases to represent data rather than generic names such as t1 or t2. However, it's crucial to consider performance implications, as nested subqueries can consume significant computational resources. To optimize performance, you need to use appropriate database indexes, which are introduced in Chapter 9.

The following story illustrates how to write complex multi-level subqueries using a food delivery platform data model. This practical business scenario shows how multi-level subqueries work.

The Second Story: A Food Delivery Platform

Ginnifer, a data analyst at a food delivery platform, faced a challenging task when the marketing team wanted to identify high-value customers for a loyalty program. She needed to find customers who spent above average in their respective cities, but only those who showed consistent ordering behavior.

194

Using Tables 7-22, 7-23, and 7-24, she wants to determine the average total amount spent by customers in each city, find customers who have spent more than that average in their respective city, and filter out customers who have placed fewer than two orders.

Table 7-22. *The Customers Table*

customer_id	Name	City
1	Alice Johnson	New York
2	Bob Smith	Chicago
3	Charlie Davis	Boston
4	David Brown	New York
5	Eva White	Chicago
6	Frank Green	Boston
7	Grace Black	New York
8	Hannah Blue	Chicago
9	Ivy Red	New York
10	Jack Gray	Chicago
11	Liam Yellow	Boston
12	Mia Purple	New York
13	Noah Orange	Chicago
14	Olivia Pink	Boston
15	Paul Silver	New York
16	Quinn Gold	Chicago

Table 7-23. *The Orders Table*

order_id	customer_id	restaurant_id	total_amount	order_date
1001	1	101	50	2025-01-01
1002	1	102	20	2025-01-02
1003	2	101	45	2025-01-03
1004	3	103	35	2025-01-04
1005	3	102	55	2025-01-05
1006	4	101	70	2025-01-06
1007	5	102	30	2025-01-07
1008	5	101	60	2025-01-08
1009	5	102	40	2025-01-09
1010	6	103	25	2025-01-10
1011	6	102	45	2025-01-11
1012	7	101	90	2025-01-12
1013	7	102	50	2025-01-13
1014	8	101	80	2025-01-14
1015	8	102	20	2025-01-15
1016	9	101	100	2025-01-16
1017	9	102	60	2025-01-17
1018	10	101	90	2025-01-18
1019	10	102	70	2025-01-19
1020	11	103	55	2025-01-20
1021	11	102	45	2025-01-21
1022	11	101	120	2025-01-22
1023	11	102	80	2025-01-23

Table 7-24. *The Restaurants Table*

restaurant_id	Name	City
101	Pizza Palace	New York
102	Sushi Spot	Chicago
103	Burger Barn	Boston

Sarah's business question is to identify customers who have spent more than the average amount in their city, but only if they have placed more than two orders.

As a first step, she breaks down the problem into the following subquestions:

- Find the average total amount spent by customers in each city.

- Find customers who have spent more than the average in the city where they live.

- Filter out customers who have placed fewer than two orders.

Multiple subqueries will be run in order to resolve each of these questions:

- The innermost subquery will calculate the average total amount spent by customers in each city.

- The middle subquery will identify customers who meet the condition.

- The outer query will filter customers based on their order count.

The following is a step-by-step breakdown of the innermost subquery:

```
SELECT c.city, AVG(o.total_amount) AS avg_city_spending
FROM customers c
JOIN orders o ON c.customer_id = o.customer_id
GROUP BY c.city;
```

The innermost subquery calculates the average total amount spent by customers in each city. This involves joining the Customers and Orders tables and grouping by the city. As shown in Table 7-25, this query calculates the average total amount spent by customers in each city.

Table 7-25. *The Average Total Amount Spent*
by Customers in Each City

City	total_amount
New York	70
Chicago	45
Boston	90

The middle subquery uses the innermost query result to filter customers who have spent more than the average in their respective cities. The middle subquery is written as follows:

```
SELECT c.customer_id, c.name, c.city, SUM(o.total_amount) AS total_spent
FROM customers c
JOIN orders o ON c.customer_id = o.customer_id
GROUP BY c.customer_id, c.name, c.city
HAVING SUM(o.total_amount) > (
SELECT AVG(o2.total_amount)
FROM customers c2
JOIN orders o2 ON c2.customer_id = o2.customer_id
WHERE c2.city = c.city GROUP BY c2.city );
```

This query identifies customers who have spent more than the average amount in their respective cities. It works by joining the Customers and Orders tables to calculate each customer's total spending. It groups results by customer details. The subquery calculates the average spending per city by joining the same tables but grouping by city only. The HAVING clause filters the customers whose total spending exceeds their city's average spending.

Note   The HAVING clause is a powerful SQL feature used to filter the results of GROUP BY operations based on aggregate conditions. While the WHERE clause filters individual rows before they're grouped, HAVING filters groups after aggregation has occurred. When you need conditions based on calculations such as SUM(), COUNT(), AVG(), MAX(), or MIN(), HAVING is essential.

The correlation between the main query and subquery is established by the `c2.city = c.city` condition, which ensures that each customer is compared only against the average from their own city. This makes the query a correlated subquery that executes once for each customer group in the outer query.

This query calculates the average total amount for each city. Lastly, the outer query is as follows:

```
SELECT c.customer_id, c.name, c.city, SUM(o.total_amount) AS total_spent,
COUNT(o.order_id) AS order_count
FROM customers c
JOIN orders o ON c.customer_id = o.customer_id
GROUP BY c.customer_id, c.name, c.city
HAVING COUNT(o.order_id) > 2 AND SUM(o.total_amount) > (
SELECT AVG(o2.total_amount)
FROM customers c2
JOIN orders o2 ON c2.customer_id = o2.customer_id
WHERE c2.city = c.city
GROUP BY c2.city );
```

This query ensures that only customers who have placed more than two orders—`COUNT(o.order_id) > 2`—and spent more than the average in their city are included in the result. The outer query retrieves customers who have spent more than the average in their city and have placed more than two orders.

When the levels are combined, a multi-level query is written in a way that provides the final answer to Sarah's question:

```
SELECT c.customer_id, c.name, c.city, SUM(o.total_amount) AS total_spent,
COUNT(o.order_id) AS order_count
FROM customers c
JOIN orders o ON c.customer_id = o.customer_id
GROUP BY c.customer_id, c.name, c.city
HAVING COUNT(o.order_id) > 2 AND SUM(o.total_amount) > (
SELECT AVG(o2.total_amount)
FROM customers c2
JOIN orders o2 ON c2.customer_id = o2.customer_id
WHERE c2.city = c.city
GROUP BY c2.city );
```

This SQL query identifies high-value customers in a food delivery platform. Total spending per customer is calculated by joining the Customers and Orders tables. In the nested subquery, orders are grouped by city to calculate the average total spending. Using the HAVING clause, the main query compares each customer's total spending with the average for their city, but only includes customers with more than two orders. In the final output, the customer's name, city, and their total spending amount are displayed for those who spent more than twice in their city. See Table 7-26.

Table 7-26. *Sarah's Business Question Final Answer*

customer_id	Name	City	total_spent	order_count
11	Liam Yellow	Boston	300	4
5	Eva White	Chicago	130	3

To improve the readability of the multi-level nested query, Sarah uses CTEs. A CTE is a temporary result set defined by the WITH clauses that can be referenced within the main query. By dividing complex logic into smaller, named blocks, SQL statements can be read, debugged, and maintained more easily. Using CTEs, Sarah can separate intermediate steps, clarify query flow, and avoid deeply nested subqueries.

```
WITH city_spending AS (
SELECT city, AVG(total_customer_spent) AS avg_city_spending
FROM (
SELECT c.customer_id, c.city, SUM(o.total_amount) AS total_customer_spent
FROM customers c
JOIN orders o ON c.customer_id = o.customer_id
GROUP BY c.customer_id, c.city ) AS customer_totals
GROUP BY city
),
customer_spending AS (
SELECT c.customer_id, c.name, c.city, SUM(o.total_amount) AS total_spent,
COUNT(o.order_id) AS order_count
FROM customers c
JOIN orders o ON c.customer_id = o.customer_id
GROUP BY c.customer_id, c.name, c.city
HAVING COUNT(o.order_id) > 2 )
```

```
SELECT cs.customer_id, cs.name, cs.city, cs.total_spent, cs.order_count
FROM customer_spending cs
JOIN city_spending csp ON cs.city = csp.city
WHERE cs.total_spent > csp.avg_city_spending;
```

This query uses CTEs to find high-value customers in a food delivery platform. It breaks down the analysis into two clear steps: First, `city_spending` calculates the total spending for each city. Second, `customer_spending` computes individual customer spending while filtering for those with more than two orders. Finally, it joins these temporary tables to identify customers whose spending exceeds their city's total. It displays their name, city, and total spent amount. This query uses CTEs for better readability and maintainability, while the previous query used nested subqueries, making it more complex.

Common Pitfalls

When working with complex nested queries and multi-level queries in SQL, developers often encounter a variety of challenges that affect performance, readability, and maintainability. This section contains a detailed overview of common pitfalls and solutions.

Poor Readability

One of the most common issues with complex nested queries is poor readability. When queries are deeply nested or overly complex, they become difficult to understand and maintain. These queries can also be challenging to debug or modify. The use of CTEs to break complex queries into smaller, more manageable parts might be one possible solution to this problem. CTEs improve readability and make it easier to isolate and troubleshoot different parts of the query.

Repeated Subquery Execution

Repeated execution of subqueries can cause performance issues, as the same computation is done multiple times during query execution.

The problem of running subqueries multiple times slows down query performance. To solve this problem, the subquery can be computed once and referenced multiple times within the query using derived tables or CTEs.

Derived tables are temporary, virtual tables that are created as a result of a subquery in a SQL query's FROM clause. These tables do not exist in the database schema but are dynamically generated at query runtime. They simplify complex queries by breaking them down into smaller, more manageable components.

The general syntax of derived tables is provided here:

```
SELECT derived_table.col1, derived_table.col2
FROM (
    SELECT col1, col2
    FROM original_table
    WHERE col3 > 100
) AS derived_table;
```

Too Many Subqueries Instead of Joins

Using too many nested subqueries instead of JOINs can make queries unnecessarily complex and inefficient. Using several nested subqueries instead of JOINs can result in slower and more complicated queries. In order to improve performance and simplify the query structure, it is best to replace subqueries with JOINs where possible.

Returning Too Much Data

Subqueries that return large result sets can negatively impact query performance because it slows down query execution. It is possible to add filters and SQL WHERE clauses to subqueries to limit the amount of data retrieved.

Forgetting to Use Aliases

Forgetting to use aliases in nested queries can result in ambiguous column references, making queries more difficult to read and maintain. Ambiguous column references in nested queries can also cause confusion and errors. To prevent this problem, use table aliases to clarify which table each column belongs to.

Table 7-27 summarizes these common pitfalls and their solutions.

Table 7-27. *Common Pitfalls and Solutions When Writing Complex Nested Queries and Multi-Level Queries*

Pitfall	Problem	Solution
Poor readability	Queries become hard to read and maintain	Use CTEs for better readability
Repeated subquery execution	Subqueries run multiple times, slowing performance	Use derived tables or CTEs
Too many subqueries instead of joins	Nested subqueries when a JOIN would suffice	Replace subqueries with JOINs
Returning too much data	Subqueries return large result sets	Add filters to subqueries
Forgetting to use aliases	Ambiguous column references in nested queries	Use table aliases to clarify

Summary

This chapter explored the concept of dynamic dialogues with subqueries, focusing on nested and multi-level queries. It highlighted how these queries can enhance SQL's flexibility by allowing complex data manipulations within a single statement. However, the chapter also discussed common pitfalls, such as poor readability, performance issues from repeated subquery executions, and challenges with large result sets. By applying best practices like CTEs and table aliases, analysts can optimize query performance and maintainability. These techniques contribute to writing more efficient, readable, and maintainable SQL code, ensuring better data analysis and decision-making outcomes.

Key Points

- Dynamic dialogues in SQL involve using subqueries to create more flexible and adaptive queries that respond to varying data inputs. These dialogues allow SQL queries to be constructed in a modular fashion, adding flexibility to SQL queries by enabling complex data manipulations within a single query.

- Multi-level nested queries provide powerful ways to solve complex problems by breaking queries into hierarchical structures. These queries enable SQL users to handle multi-step data transformations within a single statement, reducing the need for multiple queries.

- Using CTEs can improve the readability and maintainability of complex queries by breaking them into smaller, more manageable parts.

- Simplifying queries by replacing unnecessary nested subqueries with JOINs enhances both performance and clarity.

- Applying filters within subqueries reduces the amount of data processed, improving query efficiency and focusing on relevant data.

Key Takeaways

- **Dynamic dialogues with subqueries**: Use subqueries to create adaptive queries that respond dynamically to varying data inputs. This approach enhances modularity, making queries more flexible and reusable in different scenarios.

- **Multi-level nested queries**: The use of multi-level nested queries can be employed in order to handle complex data transformations within a single query. This method enables the execution of step-by-step processes without needing separate queries, improving both performance and maintainability.

Looking Ahead

The next chapter, "Conditional Logic in Data Plotting," looks at CASE statements for dynamic data outputs and explains how to customize data stories based on complex analysis of complex datasets.

Test Your Skills

A public library in a city with an extensive book collection and a large number of users employs Jack as a data analyst. Library administration is interested in improving user experience by understanding borrowing patterns, popular titles, and member activity. Jack aims to use dynamic dialogues with subqueries, nested queries, and multi-level queries to extract insights from the library database. Table 7-28 shows the BookLoans table; Table 7-29 shows the Borrowers table; and Table 7-30 shows the Books table.

Table 7-28. *The BookLoans Table*

Column	Description
loan_id	Unique identifier for each book loan
member_id	Unique identifier for each member
book_title	Title of the borrowed book
borrow_date	Date when the book was borrowed
return_date	Date when the book was returned
loan_duration	Duration in days of the loan (return_date - borrow_date)
borrow_count	Number of times the book has been borrowed

Table 7-29. *The Members Table*

Column	Description
member_id	Unique identifier for each member (primary key)
name	Full name of the member
membership_type	Type of membership (e.g., Regular, Premium)
signup_date	Date the member joined the library

Table 7-30. *The Books Table*

Column	Description
book_id	Unique identifier for each book (primary key)
book_title	Title of the book
author	Author of the book
genre	Genre of the book
published_year	Year the book was published

1. Determine which books are most frequently borrowed. Try writing a query using a subquery to find the top five most borrowed books from the BookLoans table.

2. Determine which members have the longest average loan duration. Try using a multi-level query to calculate the average loan duration for each member and identify the top three members with the longest average loan duration.

3. Analyze the borrowing patterns of recent times. Try writing a query using a subquery to list the ten most recent loans, including member_id and book_title, sorted by borrow_date.

4. Conduct an analysis of borrowing frequency. Try to use a CTE to identify members who have borrowed more than 20 books and calculate their average loan duration.

5. Compare the borrower's behavior over two time periods. Try to use a nested query to compare the number of books borrowed in the first half of the year with the second half.

CHAPTER 8

Conditional Logic in Data Plotting

In this chapter, you learn how SQL's conditional logic can transform data analysis and visualization workflows. You learn about conditional logic in SQL, categorize data, apply dynamic filtering to improve plot relevance for enhanced visualization, create color-coding data for visualizations, aggregate data using conditional expressions, and handle missing data in visualizations.

The focus in this chapter is not on how you visualize or plot data, but rather on the crucial process of preparing data before visualization using SQL. The primary contribution of this chapter is to explain how SQL can be used to manipulate, clean, and structure data before it reaches the visualization stage. SQL itself does not generate plots or graphical representations; instead, it serves as a powerful tool for transforming raw data into a structured format suitable for analysis. While other programming languages and tools handle data visualization, SQL ensures that the underlying data is properly processed, making it ready for effective and meaningful representation. However, some database management systems, such as `pgAdmin` for PostgreSQL, offer limited visualization tools for displaying query results graphically. But these are not part of standard SQL itself—they are features of the database's user interface. In order to plot data, SQL can be combined with external tools or programming languages, such as Python (Matplotlib, Seaborn, Plotly, Pandas), R (ggplot2, tidyverse), or BI tools (Tableau, Power BI, Looker, Metabase).

Introduction

SQL conditional logic can be implemented using expressions such as `CASE`, `IF`, and conditional operators like <, >, =, and so on. Using these constructs, developers can dynamically transform or filter data based on specific criteria. This makes SQL queries

© Hamed Tabrizchi 2025
H. Tabrizchi, *Narrative SQL*, https://doi.org/10.1007/979-8-8688-1560-7_8

more flexible and tailored to the needs of complex analytical tasks by allowing them to manipulate datasets to meet particular requirements. For instance, conditional logic can categorize data into groups, create dynamic columns based on conditions, and perform context-sensitive calculations within the query itself.

By using SQL conditional expressions for dynamic plotting, you can enhance your data analysis because they allow dynamic and context-aware plotting of datasets. As part of data preparation, it is possible to use CASE statements to create conditional labels for categorical axes in plots, such as grouping age ranges or income levels. To perform context-aware filtering, one option is to use conditional filtering to include or exclude data points based on analysis-specific thresholds or trends, ensuring that the plots reflect meaningful insights. Also, for derived metrics, it is possible to calculate new metrics directly in the query to highlight specific trends, such as defining high- or low-performance thresholds dynamically.

Table 8-1 summarizes the common visualization plot types and the PostgreSQL queries used to prepare the data. The table includes the plot type, its use case, and an example query that prepares data in PostgreSQL. Each example query in this table uses the `sales_data` table as its base and has an explanation column to explain its purpose. The `sales_data` table contains the following columns: `date` (DATE), which represents the transaction date; `region` (VARCHAR), indicating the sales region; `product` (VARCHAR), specifying the product name; `category` (VARCHAR), defining the product category; `sales` (NUMERIC), denoting the sales amount; and `profit` (NUMERIC), representing the profit amount.

Table 8-1. *Common Visualization Plot Types and the PostgreSQL Queries Used to Prepare the Data*

Plot Type	Plot Use Case	Example PostgreSQL Query	Explanation

Bar chart	Compare categorical data, for instance, sales by region	`SELECT region,`
		`SUM(sales) AS total_`
		`sales FROM sales_data`
		`GROUP BY region;`
			Groups sales by region to show total sales per region.

Line chart	Show trends over time, for instance, stock prices	`SELECT date, SUM(sales)`
		`AS daily_sales FROM`
		`sales_data GROUP BY`
		`date ORDER BY date;`
			Aggregates daily sales to show trends over time.

Scatterplot	Visualize relationships between two numeric variables, for instance, height vs. weight	`SELECT sales, profit` `FROM sales_data;`	Extracts sales and profit values to visualize their relationship.

Pie chart	Show proportions or percentages of categories, for instance, market share	`SELECT category,`
		`SUM(sales) AS total_`
		`sales FROM sales_data`
		`GROUP BY category;`
			Groups sales by category to show proportions of total sales.

Histogram	Show frequency distributions, for instance, age distribution	`SELECT FLOOR(sales`
		`/ 100) * 100 AS`
		`sales_range, COUNT(*)`
		`AS frequency FROM`
		`sales_data GROUP BY`
		`sales_range ORDER BY`
		`sales_range;`
			Buckets sales into ranges, for instance 0–100, 100–200, and counts transactions in each range for a distribution.

(continued)

Table 8-1. (*continued*)

Plot Type	Plot Use Case	Example PostgreSQL Query	Explanation
Box plot	Display data distribution and outliers, for example, test scores by class	`SELECT region, sales FROM sales_data;`	Retrieves sales per region to analyze distribution and identify outliers.
Heatmap	Show patterns across two variables, for instance, sales by day and time	`SELECT date, region, SUM(sales) AS total_ sales FROM sales_data GROUP BY date, region;`	Groups sales by date and region to visualize patterns in a heatmap.
Stacked bar chart	Compare proportions across categories, for example, sales by product and region	`SELECT region, category, SUM(sales) AS total_sales FROM sales_ data GROUP BY region, category;`	Groups sales by region and category to show category-wise contributions in each region.
Area chart	Show cumulative trends over time, for instance, cumulative sales	`SELECT date, SUM(sales) OVER (ORDER BY date) AS cumulative_sales FROM sales_data;`	Calculates cumulative sales over time to show trends in total sales growth.
Bubble chart	Represent three dimensions, for example, profit, sales, and region	`SELECT region, SUM(sales) AS total_ sales, SUM(profit) AS total_profit FROM sales_data GROUP BY region;`	Summarizes sales and profit by region to represent three dimensions, for example, bubble size = sales volume.
Radar chart	Compare multiple variables for categories, for instance, performance metrics	`SELECT category, AVG(sales) AS avg_ sales, AVG(profit) AS avg_profit FROM sales_ data GROUP BY category;`	Aggregates average sales and profit by category for multi-dimensional comparisons.

(*continued*)

Table 8-1. (*continued*)

Plot Type	Plot Use Case	Example PostgreSQL Query	Explanation
Waterfall chart	Visualize changes over time or between categories, for instance, profit margins	`SELECT product, SUM(sales) AS total_ sales FROM sales_data GROUP BY product ORDER BY total_sales DESC;`	Ranks products by sales for visualizing incremental changes between top-selling products.
Gantt chart	Track project schedules, for instance, start and end dates for tasks	`SELECT product, MIN(date) AS start_ date, MAX(date) AS end_ date FROM sales_data GROUP BY product;`	Calculates start and end dates of sales for each product for timeline-based visualization.
Treemap	Show hierarchical data, for instance, sales by product category and subcategory	`SELECT category, product, SUM(sales) AS total_sales FROM sales_ data GROUP BY category, product;`	Groups sales by category and product to represent hierarchical data.
Donut chart	Similar to a pie chart but with a central cut-out, for example, profit share	`SELECT category, SUM(profit) AS total_ profit FROM sales_data GROUP BY category;`	Groups profit by category to show proportions, similar to a pie chart but with a central cut-out.
Pareto chart	Highlight the most significant factors in data, for instance, sales by product	`SELECT product, SUM(sales) AS total_ sales FROM sales_data GROUP BY product ORDER BY total_sales DESC;`	Ranks products by sales to identify the most significant contributors, for example, the 80/20 rule.
Violin plot	Show distribution and density, for instance, salary ranges	`SELECT category, sales FROM sales_data;`	Extracts sales data for each category to show distribution density and variability.

(*continued*)

Table 8-1. (*continued*)

Plot Type	Plot Use Case	Example PostgreSQL Query	Explanation
Chord diagram	Visualize relationships between entities, for instance, trade flows	`SELECT region AS source, category AS target, SUM(sales) AS value FROM sales_data GROUP BY region, category;`	Maps relationships between regions (source) and categories (target) based on sales.
Network graph	Show connections between nodes, for example, social networks	`SELECT region AS source, product AS target FROM sales_ data GROUP BY region, product;`	Highlights connections between regions and products for relationship analysis.
Funnel chart	Represent stages in a process, for instance, sales pipeline	`SELECT category, COUNT(*) AS total_ transactions FROM sales_data GROUP BY category ORDER BY total_transactions DESC;`	Tracks the number of transactions for each category to represent stages in a process, for example, sales funnel.

Understanding Conditional Logic in SQL

SQL conditional logic allows you to control query output based on specific conditions. As a result, it is crucial for categorizing data, handling missing values, avoiding errors such as division by zero, and returning different results depending on the conditions. CASE, NULLIF, and COALESCE are three of the most commonly used conditional expressions in PostgreSQL. In the form of an `if-else` statement, CASE expression is the most powerful conditional expression. The NULLIF function returns NULL if two values are equal, otherwise it returns the first value. From a list, the COALESCE function returns the first value that is not null.

The CASE Statement

A CASE statement allows a query to be processed in a way similar to an `if-else` statement. Consider an example that categorizes sales amounts. In Table 8-2, the `order_id` and `amount` are listed.

Table 8-2. *The Orders Table*

order_id	Amount
101	1200
102	700
103	300

The following query retrieves `order_id` and `amount` from an `Orders` table and adds a new column called `category` using a CASE statement, which is shown in Table 8-3. The `category` column classifies order amounts into `High` ($\geq$ 1000), `Medium` ($\geq$ 500), and `Low` (< $500), based on their monetary value. This allows for quick visualization and analysis of order values by creating a simple categorical breakdown without modifying the original table structure.

```
SELECT order_id, amount,
    CASE
        WHEN amount >= 1000 THEN 'High'
        WHEN amount >= 500 THEN 'Medium'
        ELSE 'Low'
    END AS category
FROM orders;
```

The results of this query are shown in Table 8-3.

Table 8-3. *Categorizing order_id and Amount from an Orders Table*

order_id	Amount	Category
101	1200	High
102	700	Medium
103	300	Low

This next query example provides a solution to a more complicated business intelligence question. The query aims to show how to categorize and understand product categories based on their sales performance.

This query must answer the following sub-questions:

- How many product categories are there?

- What is the sales volume for each product category?

- How can you classify product categories into meaningful performance tiers?

- What are the averages and total sales for each category?

Here, the role of the CASE statement is crucial because it creates a meaningful categorization of sales volumes, transforms raw numerical data into actionable insights, and allows for quick visual and analytical understanding of product category performance.

Given the data in Table 8-4, the `sales_data` table, it is possible to identify top-performing product categories, understand the distribution of sales across different product lines, and make informed decisions about inventory, marketing, and strategic planning.

Table 8-4. *The sales_data Table*

product_category	sale_amount	sale_date
Electronics	1500	2024-01-15
Clothing	450	2024-01-16
Electronics	2300	2024-01-17
Books	750	2024-01-18
Clothing	1100	2024-01-19
Electronics	3200	2024-01-20
Books	250	2024-01-21
Clothing	600	2024-01-22
Electronics	4500	2024-01-23
Books	890	2024-01-24
Clothing	1750	2024-01-25
Electronics	5600	2024-01-26
Books	330	2024-01-27
Clothing	880	2024-01-28
Electronics	6700	2024-01-29

The following query demonstrates a structured approach to preparing sales data. It uses a subquery to calculate total sales per product category, providing a broad overview of sales distribution. A CASE statement categorizes these totals into sales volume tiers based on predefined thresholds, allowing flexible classification. Aggregate functions compute key metrics such as the number of sales, the average sales per category, and total sales. This makes the resulting dataset suitable for visualization and further analysis. By combining dynamic categorization and ordering by total sales, the query presents insights in a clear and meaningful format:

```
SELECT
    product_category,
    CASE
        WHEN SUM(sale_amount) >= 10000 THEN 'High Volume'
```

```
        WHEN SUM(sale_amount) >= 5000 THEN 'Medium Volume'
        ELSE 'Low Volume'
    END AS sales_volume_category,
    COUNT(*) AS number_of_sales,
    AVG(sale_amount) AS average_sale,
    SUM(sale_amount) AS total_sales
FROM sales_data
GROUP BY product_category
ORDER BY total_sales DESC;
```

The results of this query are shown in Table 8-5.

Table 8-5. *The Structured Data Table (After Query Execution)*

product_category	sales_volume_category	number_of_sales	average_sale	total_sales
Electronics	High Volume	6	3966.66	23800.00
Clothing	Low Volume	5	956.00	4780.00
Books	Low Volume	4	555.00	2220.00

This query efficiently prepares structured data that can be used for meaningful insights. The computed fields—such as sales_volume_category, number_of_sales, average_sale, and total_sales—are key elements for data visualization. In the next stage, this data can be illustrated using various chart types: a *bar chart* can compare absolute values between categories; a *pie chart* shows the proportion of total sales per category, and a *scatterplot* can extend the data to explore relationships, such as between sales volume and frequency.

NULLIF

The NULLIF function prevents errors by returning NULL if two values are equal. NULLIF is a powerful SQL function that prevents division by zero errors and handles null values during data analysis. In essence, it allows you to replace a specific value with NULL, which is crucial when performing calculations like percentages or ratios where zero could cause computational issues. Using NULLIF, data analysts and database professionals can create reliable, error-resistant queries for reporting, plotting, and statistical analysis by converting potentially problematic zero values to NULL. In data transformation, this provides a clean, robust way to deal with edge cases.

Note   In SQL, certain common scenarios can lead to errors or unexpected behavior if they aren't handled carefully. These include division by zero, distinguishing between empty strings and NULL values, date range boundaries, case sensitivity in string comparisons, numerical precision loss, and aggregation over empty datasets. These are not rare edge cases; they are critical considerations in writing robust SQL queries and should be explicitly handled during data processing.

In the case of zero targets or missing data, for instance, NULLIF can be used to analyze the following employee performance metrics data. Table 8-6 shows employee performance metrics, including sales figures, targets, training hours, and completed projects. This gives a structured view of every employee's sales achievements in relation to their targets.

Table 8-6. *The performance_metrics Data Table (Employee Performance Metrics)*

emp_id	Name	Sales	Target	training_hours	projects_completed
101	Alice	50000	45000	20	5
102	Bob	30000	0	15	3
103	Charlie	75000	60000	0	8
104	Diana	45000	40000	25	0
105	Eve	0	35000	30	4
106	Frank	85000	80000	10	7
107	Grace	25000	30000	0	2
108	Henry	55000	0	40	6
109	Ivy	65000	50000	35	0
110	Jack	40000	45000	0	5

The following query evaluates employee performance by calculating key metrics such as target achievement, productivity ratio, and performance variance. Using a CTE, it first computes target achievement as a percentage while handling division by zero with `NULLIF(target, 0)`. The productivity ratio is derived from completed projects relative to training hours, and performance variance measures deviation from sales targets. The main query categorizes employees into performance statuses and efficiency ratings based on predefined thresholds. NULL values are handled effectively, ensuring robust data analysis. In this query, the results are sorted by target achievement, placing high-performing employees at the top.

```
WITH performance_metrics AS (
    SELECT
        emp_id,
        name,
        ROUND(
            sales::numeric / NULLIF(target, 0) * 100,
            2
        ) AS target_achievement,
        ROUND(
            projects_completed::numeric / NULLIF(training_hours, 0) * 100,
            2
        ) AS productivity_ratio,
        CASE
            WHEN sales = 0 THEN NULL
            WHEN target = 0 THEN NULL
            ELSE ROUND((sales - target)::numeric / target * 100, 2)
        END AS performance_variance
    FROM employee_metrics
)
SELECT
    pm.*,
    CASE
        WHEN target_achievement IS NULL THEN 'Invalid Target'
        WHEN target_achievement >= 100 THEN 'Exceeded'
        WHEN target_achievement >= 80 THEN 'On Track'
        ELSE 'Below Target'
```

```
    END AS performance_status,
    CASE
        WHEN productivity_ratio IS NULL THEN 'No Training Data'
        WHEN productivity_ratio > 20 THEN 'High Efficiency'
        WHEN productivity_ratio > 10 THEN 'Moderate Efficiency'
        ELSE 'Needs Improvement'
    END AS efficiency_rating
FROM performance_metrics pm
ORDER BY target_achievement DESC NULLS LAST;
```

This query uses the NULLIF function to handle potential division errors and ensure accurate calculations, which makes it a good example of how NULLIF plays an important role. By applying NULLIF(target, 0), the query prevents division by zero when computing target achievement. NULLIF(training_hours, 0) ensures productivity calculations remain valid even without training hours. Additionally, NULLIF is combined with CASE statements to create meaningful categorizations. This allows the query to manage various zero and NULL scenarios in a single analysis.

In this query, ::numeric casting is used to ensure decimal division and precise rounding, especially in cases where the original data type might be integers. For example, without casting, dividing two integers could result in integer division, for instance 1/2 = 0 instead of 0.5. If the columns involved are already of a NUMERIC or DECIMAL type, the castings can be safely removed. However, if the schema uses INTEGER or TEXT, the explicit cast guarantees consistent behavior during division and rounding operations.

In this query, a WITH clause is used to create a CTE named performance_metrics. This CTE calculates several performance-related metrics for employees based on the employee_metrics table. After defining the CTE, the main query selects from it and adds columns (performance_status and efficiency_rating) based on the calculated metrics.

Note   As mentioned in earlier chapters, a WITH clause, also known as a CTE (common table expression), allows you to define a temporary result set that you can reference within a SELECT, INSERT, UPDATE, or DELETE statement. It is often used to simplify complex queries by breaking them into smaller, more manageable parts. The basic syntax of the WITH clause is as follows:

```
WITH cte_name AS (
    -- Subquery that defines the CTE
    SELECT ...
    FROM ...
    WHERE ...
)
-- Main query that references the CTE
SELECT ...
FROM cte_name
WHERE ...
```

The transformation of raw metrics into actionable insights is achieved by safely calculating target achievement percentages. This is done by computing productivity ratios without errors, determining performance variance, and categorizing both performance and efficiency levels. Moreover, the query accounts for edge cases where targets or training hours are zero, ensuring the analysis remains robust. This example highlights how NULLIF plays a crucial role in data analysis, particularly when working with real-world datasets that often include missing values or zero entries. See Table 8-7.

Table 8-7. The Result of Safely Calculating Target Achievement Percentages

emp_id	name	target_achievement	productivity_ratio	performance_variance	performance_status	efficiency_rating
106	Frank	106.25	70	6.25	Exceeded	High Efficiency
103	Charlie	125	NULL	25	Exceeded	No Training Data
109	Ivy	130	0	30	Exceeded	Needs Improvement
101	Alice	111.11	25	11.11	Exceeded	High Efficiency
104	Diana	112.5	0	12.5	Exceeded	Needs Improvement
110	Jack	88.89	NULL	-11.11	On Track	No Training Data
107	Grace	83.33	NULL	-16.67	On Track	No Training Data
105	Eve	0	13.33	-100	Below Target	Moderate Efficiency
102	Bob	NULL	20	NULL	Invalid Target	Moderate Efficiency
108	Henry	NULL	15	NULL	Invalid Target	Moderate Efficiency

Now, since NULL handling is included, this dataset is valuable for in-depth performance assessments and decision-making. Table 8-7 shows how the performance evaluation is extended by adding calculated metrics such as targets achieved, productivity ratios, performance variances, performance status, and efficiency ratings. Target achievement is measured as a percentage, ensuring that employees exceeding their goals are recognized. The productivity ratio assesses efficiency based on training hours, while performance variance captures deviations from expected performance. Employees are categorized into different performance statuses, such as Exceeded or On Track, with efficiency ratings indicating their effectiveness.

COALESCE

COALESCE in PostgreSQL is a function that returns the first non-NULL value from a list of expressions. Therefore, it is an essential component for handling NULL values and providing default values. In the case of a table that contains potentially missing data (NULLS) in columns, COALESCE allows you to substitute these NULLs with meaningful default values. The basic syntax of COALESCE in PostgreSQL is as follows:

```
COALESCE(value1, value2, ..., value_n)
```

COALESCE returns the first non-NULL value from the provided arguments list. If all values are NULL, it returns NULL. It is commonly used to handle NULL values in queries by providing a default or fallback value.

Table 8-8 shows a membership data table that must be filled in with missing expiration dates by assuming a default expiration period.

Table 8-8. *The Memberships Data Table*

member_id	member_name	membership_type	expiration_date
1	John Doe	Gold	2025-06-15
2	Jane Smith	Silver	NULL
3	Mike Johnson	Gold	2025-08-10
4	Sarah Williams	Bronze	NULL
5	David Brown	Gold	2025-07-01
6	Emily Davis	Silver	NULL
7	James Wilson	Bronze	2025-05-20
8	Laura Martinez	Gold	NULL
9	Robert Taylor	Silver	2025-09-25
10	Sophia Anderson	Bronze	NULL

The following query aims to use COALESCE to fill in missing expiration dates by assuming the default expiration period for missing values is: Gold: 2025-12-31, Silver: 2025-11-30, and Bronze: 2025-10-31.

```
SELECT
    member_id,
    member_name,
    membership_type,
    COALESCE(expiration_date,
        CASE
            WHEN membership_type = 'Gold' THEN DATE '2025-12-31'
            WHEN membership_type = 'Silver' THEN DATE '2025-11-30'
            WHEN membership_type = 'Bronze' THEN DATE '2025-10-31'
        END
    ) AS adjusted_expiration_date
FROM memberships;
```

In this case, these values were assumed, but in real-world scenarios, best strategies for missing value treatment have been mostly followed using this process:

1. Analyze patterns in missing data.

2. Considering the business context, document assumptions and methods.

3. Validate imputed values against known patterns and statistical methods with domain expertise to ensure that the imputed values make business sense and maintain data integrity.

By using COALESCE with a CASE statement, you can ensure that members with missing expiration dates receive a default date based on their membership type. This helps maintain data consistency and avoids NULL values in reports. This query ensures that members with missing expiration dates will be provided a default date. This is based on their membership type. As shown in Table 8-9, this helps maintain data consistency and avoids NULL values in reports.

Table 8-9. *Dates of Expiration That Were Missing Have Been Filled In*

member_id	member_name	membership_type	adjusted_expiration_date
1	John Doe	Gold	2025-06-15
2	Jane Smith	Silver	2025-11-30
3	Mike Johnson	Gold	2025-08-10
4	Sarah Williams	Bronze	2025-10-31
5	David Brown	Gold	2025-07-01
6	Emily Davis	Silver	2025-11-30
7	James Wilson	Bronze	2025-05-20
8	Laura Martinez	Gold	2025-12-31
9	Robert Taylor	Silver	2025-09-25
10	Sophia Anderson	Bronze	2025-10-31

The First Story: The Hospital's Analytical Story

Sofia, a data analyst at a large metropolitan hospital, has been tasked with preparing visualizations for a key meeting. The raw hospital data is messy, with missing values, outliers, and complex relationships between variables. As illustrated in Table 8-10, a `hospital_admissions` table has 20 rows and several NULL values. Despite this, Sofia continues to make clean, insightful plots with SQL and her analytical mind.

Table 8-10. *The hospital_admissions Table*

patient_id	Age	diagnosis_code	admission_date	discharge_date
1	12	150	2024-03-01	2024-03-03
2	45	120	2024-03-02	2024-03-05
3	67	180	2024-03-03	2024-03-10
4	8	90	2024-03-04	2024-03-06
5	32	110	2024-03-05	2024-03-08
6	55	130	2024-03-06	2024-03-12
7	70	140	2024-03-07	NULL
8	25	160	2024-03-08	2024-03-11
9	40	170	2024-03-09	2024-03-16
10	60	190	2024-03-10	2024-03-20
11	15	NULL	2024-03-11	2024-03-13
12	50	200	2024-03-12	NULL
13	28	105	2024-03-13	2024-03-15
14	72	NULL	2024-03-14	2024-03-21
15	10	115	2024-03-15	NULL
16	35	125	2024-03-16	2024-03-18
17	65	135	2024-03-17	2024-03-24
18	22	145	2024-03-18	2024-03-20
19	48	155	2024-03-19	NULL
20	58	165	2024-03-20	2024-03-27

She encounters six major challenges during the data-preparation process:

- How can she categorize hospital patients into age groups such as Child, Adult, and Senior, to make visualizations of admissions by age group easier to interpret?

- How can she dynamically filter the hospital data to show only patients admitted for cardiac-related issues—for instance, diagnosis codes between 100 and 199 during the year 2024?

- What is the average length of stay for patients grouped by diagnosis_ code ranges (100–125, 126–150, 151–175, and 176–199) for cardiac-related issues?

- How can she preprocess hospital data to create a stay_category column, using conditional logic to classify patient stays as Short Stay, Moderate Stay, or Long Stay, based on the number of days stayed?

- How can she calculate the total number of patients and the average length of stay for each age group (Child, Adult, Senior) to identify trends and resource needs for different demographics?

- How can she clean the hospital data for visualization by handling missing discharge dates (filling them with admission_date + 2 days) and excluding outliers where the length of the stay exceeds 365 days?

As part of the first challenge, the aim is to categorize hospital patients by age, including their age data values in a hospital_admissions data table. Sofia needs to categorize patients into age groups (Child, Adult, Senior) to better visualize admissions by age category.

```
SELECT
    patient_id,
    age,
    CASE
        WHEN age < 18 THEN 'Child'
        WHEN age BETWEEN 18 AND 64 THEN 'Adult'
        WHEN age >= 65 THEN 'Senior'
        ELSE 'Unknown'
    END AS age_group
```

```
FROM hospital_admissions
ORDER BY age_group;
```

This query creates an age_group column using conditional logic. Categorizing data like this improves clarity in visualizations by grouping similar data points together. In this query, the CASE statement evaluates the age column and assigns a category based on the value. It categorizes individuals as Child if the age is less than 18, Adult if their age is between 18 and 64, and Senior if their age is 65 or older. For any unexpected values, such as NULL, the Unknown category is assigned.

In this query, the CASE statement classifies patients into age groups based on their age. The BETWEEN operator is used to define the Adult category. It's important to note that BETWEEN is *inclusive,* meaning that the boundary values, 18 and 64, are included. So, a patient who is exactly 18 or exactly 64 years old will fall into the Adult category. This ensures a clear, non-overlapping classification across all age ranges.

As shown in Table 8-11, the results are sorted by the age_group column using the ORDER BY clause, ensuring easier readability.

Table 8-11. *Data Categorization by age_group*

patient_id	Age	age_group
1	12	Child
4	8	Child
11	15	Child
15	10	Child
2	45	Adult
5	32	Adult
6	55	Adult
8	25	Adult
9	40	Adult
10	60	Adult
13	28	Adult

(*continued*)

Table 8-11. (*continued*)

patient_id	Age	age_group
16	35	Adult
18	22	Adult
19	48	Adult
20	58	Adult
3	67	Senior
7	70	Senior
14	72	Senior
17	65	Senior

In the next challenge, decision makers want a dynamic view of admitted patients according to their diagnosis code. Sofia needs to filter data so that only cardiac-related patients with diagnosis_codes of 100 to 199 are displayed.

```
SELECT
    patient_id,
    diagnosis_code,
    admission_date,
    discharge_date
FROM hospital_admissions
WHERE diagnosis_code BETWEEN 100 AND 199
  AND admission_date >= '2024-01-01'
  AND admission_date <= '2024-12-31';
```

In this query, the WHERE clause allows dynamic filtering. Here, data is narrowed to show only cardiac-related admissions in a specific year, reducing clutter and focusing the plot on the requested subset of data. The WHERE clause filters rows by selecting only those where the diagnosis_code is between 100 and 199, representing cardiac-related issues, and the admission_date falls within the year 2024. As shown in Table 8-12, the dynamic filtering ensures that only relevant data is included for analysis.

Table 8-12. *Dynamic Filtering (Cardiac-Related Issues)*

patient_id	diagnosis_code	admission_date	discharge_date
1	150	2024-03-01	2024-03-03
2	120	2024-03-02	2024-03-05
3	180	2024-03-03	2024-03-10
5	110	2024-03-05	2024-03-08
6	130	2024-03-06	2024-03-12
7	140	2024-03-07	NULL
8	160	2024-03-08	2024-03-11
9	170	2024-03-09	2024-03-16
10	190	2024-03-10	2024-03-20
13	105	2024-03-13	2024-03-15
15	115	2024-03-15	NULL
16	125	2024-03-16	2024-03-18
17	135	2024-03-17	2024-03-24
18	145	2024-03-18	2024-03-20
19	155	2024-03-19	NULL
20	165	2024-03-20	2024-03-27

In healthcare and hospital records, a diagnosis code is a standardized code used to represent a specific medical condition, diagnosis, or reason for a patient's admission to the hospital. These codes are typically part of a coding system like the International Classification of Diseases. This system is widely used in healthcare for statistical, billing, and research purposes. For example, diagnosis codes in the range 100 to 199 might correspond to cardiac-related conditions like heart disease, arrhythmia, or hypertension.

The following query calculates the average length of stay for cardiac-related issues grouped by diagnosis_code ranges (100–125, 126–150, 151–175, and 176–199).

```
SELECT
    CASE
        WHEN diagnosis_code BETWEEN 100 AND 125 THEN '100-125'
        WHEN diagnosis_code BETWEEN 126 AND 150 THEN '126-150'
        WHEN diagnosis_code BETWEEN 151 AND 175 THEN '151-175'
        WHEN diagnosis_code BETWEEN 176 AND 199 THEN '176-199'
        ELSE 'Unknown'
    END AS diagnosis_code_range,
    COUNT(patient_id) AS patient_count,
    AVG(COALESCE(discharge_date, admission_date + INTERVAL '2 days') -
    admission_date) AS avg_length_of_stay
FROM hospital_admissions
WHERE diagnosis_code BETWEEN 100 AND 199
  AND admission_date >= '2024-01-01'
  AND admission_date <= '2024-12-31'
GROUP BY diagnosis_code_range
ORDER BY diagnosis_code_range;
```

The query uses a CASE statement to group the diagnosis_code into defined ranges. The COALESCE function replaces NULL discharge_date values with admission_date plus two days to calculate the length of stay for all patients. The AVG function calculates the average length of stay for each diagnosis_code range. The COUNT function determines the number of patients in each range, providing better insights into patient distribution. The WHERE clause filters the data to include only cardiac-related admissions in 2024. The query provides valuable insights into average lengths of stay across diagnosis_code ranges, assisting decision-makers in optimizing resources and understanding trends.

Next, to create a scatterplot of patient ages against length of stay, Sofia wants to assign colors based on whether the stay was short (< 3 days), moderate (3-7 days), or long (> 7 days).

```
SELECT
    patient_id,
    age,
    (discharge_date - admission_date) AS length_of_stay,
    CASE
        WHEN (discharge_date - admission_date) < 3 THEN 'Short Stay'
```

```
      WHEN (discharge_date - admission_date) BETWEEN 3 AND 7 THEN
      'Moderate Stay'
      WHEN (discharge_date - admission_date) > 7 THEN 'Long Stay'
      ELSE 'Unknown'
   END AS stay_category
FROM hospital_admissions;
```

Conditional logic in CASE creates a stay_category column, which can be used to assign colors to plot points in visualization software (for instance, Short Stay = Blue, Moderate Stay = Green, Long Stay = Red).

In this query, the length_of_stay calculation determines the difference between the discharge_date and admission_date. A CASE statement then categorizes the length of stay as a Short Stay for stays less than three days, a Moderate Stay for stays between three and seven days, and a Long Stay for stays exceeding seven days. As shown in Table 8-13, for missing or invalid values, the Unknown category is assigned.

Table 8-13. *Results of the Length of Stay Calculation*

patient_id	Age	length_of_stay	stay_category
1	12	2	Short Stay
2	45	3	Moderate Stay
3	67	7	Moderate Stay
4	8	2	Short Stay
5	32	3	Moderate Stay
6	55	6	Moderate Stay
7	70	NULL	Unknown
8	25	3	Moderate Stay
9	40	7	Moderate Stay
10	60	10	Long Stay
11	15	2	Short Stay
12	50	NULL	Unknown

(*continued*)

Table 8-13. (*continued*)

patient_id	Age	length_of_stay	stay_category
13	28	2	Short Stay
14	72	7	Moderate Stay
15	10	NULL	Unknown
16	35	2	Short Stay
17	65	7	Moderate Stay
18	22	2	Short Stay
19	48	NULL	Unknown
20	58	7	Moderate Stay

Next, the hospital administrators want to know the average length of stay for each age group (Child, Adult, Senior) to plan resources more effectively.

```
SELECT
    CASE
        WHEN age < 18 THEN 'Child'
        WHEN age BETWEEN 18 AND 64 THEN 'Adult'
        WHEN age >= 65 THEN 'Senior'
        ELSE 'Unknown'
    END AS age_group,
    COUNT(patient_id) AS total_patients,
    AVG(discharge_date - admission_date) AS avg_length_of_stay
FROM hospital_admissions
GROUP BY age_group
ORDER BY age_group;
```

In this query, using conditional logic within the aggregation provides meaningful insights, such as the average length of stay by age group. This type of summarization is key for creating bar charts or line plots showing trends. Here, the CASE statement groups patients into categories of Child, Adult, and Senior. Aggregation functions are then applied, with COUNT(patient_id) used to count the number of patients in each group and AVG(`discharge_date - admission_date`) calculating the average length of stay for each group. As shown in Table 8-14, the results are grouped by age_group using the GROUP BY clause.

Table 8-14. *The Average Length of Stay Categorized by Age Group*

age_group	total_patients	avg_length_of_stay
Child	4	2
Adult	12	4.5
Senior	4	7

The next challenge is that some patients' discharge dates are missing, while others have extremely long lengths of stay (e.g., outliers caused by data entry errors). Sofia needs to clean this data for plotting. Missing values should be filled with a default value of `admission_date` plus two days, and any length of stay over 365 days should be excluded.

```
SELECT
    patient_id,
    age,
    admission_date,
    COALESCE(discharge_date, admission_date + INTERVAL '2 days') AS clean_
    discharge_date,
    GREATEST(0, EXTRACT(DAY FROM COALESCE(discharge_date, admission_date +
    INTERVAL '2 days') - admission_date)) AS clean_length_of_stay
FROM hospital_admissions
WHERE EXTRACT(DAY FROM COALESCE(discharge_date, admission_date + INTERVAL
'2 days') - admission_date) <= 365;
```

This query uses COALESCE to handle missing values and ensures only clean, valid data is used in visualizations, as shown in Table 8-15. Outliers are excluded by applying a filter on length_of_stay. The COALESCE function replaces NULL discharge dates with admission_date plus two days. The GREATEST function ensures that the length of stay is at least 0 days. Additionally, the WHERE clause excludes rows where the length of stay exceeds 365 days, filtering out outliers.

Table 8-15. *Cleaning the Data for Plotting and Handling Missing Values Filled with a Default Value*

patient_id	Age	admission_date	clean_discharge_date	clean_length_of_stay
1	12	2024-03-01	2024-03-03	2
2	45	2024-03-02	2024-03-05	3
3	67	2024-03-03	2024-03-10	7
4	8	2024-03-04	2024-03-06	2
5	32	2024-03-05	2024-03-08	3
6	55	2024-03-06	2024-03-12	6
7	70	2024-03-07	2024-03-09	2
8	25	2024-03-08	2024-03-11	3
9	40	2024-03-09	2024-03-16	7
10	60	2024-03-10	2024-03-20	10
11	15	2024-03-11	2024-03-13	2
12	50	2024-03-12	2024-03-14	2
13	28	2024-03-13	2024-03-15	2
14	72	2024-03-14	2024-03-21	7
15	10	2024-03-15	2024-03-17	2
16	35	2024-03-16	2024-03-18	2
17	65	2024-03-17	2024-03-24	7
18	22	2024-03-18	2024-03-20	2
19	48	2024-03-19	2024-03-21	2
20	58	2024-03-20	2024-03-27	7

Summary

This chapter highlighted the power of SQL's conditional logic in enhancing data analysis and visualization workflows. You explored techniques to categorize data, apply dynamic filtering, and handle missing or outlier values for more relevant visualizations.

Conditional logic enables context-aware plotting by dynamically including or excluding data points based on specific thresholds or trends. CASE statements allow conditional labels for plots, such as grouping age ranges or income levels. SQL also simplifies the calculation of derived metrics, making it easier to identify high- or low-performance trends. These techniques collectively improve the accuracy, clarity, and insight of data visualizations.

Key Points

- Conditional logic in SQL enhances data analysis by allowing dynamic and context-aware plotting, enabling more meaningful visualizations.

- Using CASE statements, data can be categorized or conditionally labeled. For example, grouping age ranges or income levels for a clearer representation.

- Dynamic filtering ensures plots focus on relevant data points by including or excluding data based on thresholds or trends.

- SQL's flexibility handles missing and outlier data, improving visualization accuracy and insights.

- By using SQL conditional logic, workflows become more adaptive, ensuring that visualizations reflect valuable and actionable insights.

Key Takeaways

- **Dynamic data categorization**: Use CASE statements to group or label data dynamically for clearer and more meaningful visualizations.

- **Context-aware filtering**: Apply conditional filtering to include or exclude data points based on specific thresholds or trends, ensuring that plots focus on relevant insights.

- **Derived metrics for insights**: Use conditional expressions to calculate relevant metrics directly within queries, enabling the identification of patterns like high- or low-performance thresholds.

- **Improved visualization accuracy**: Handle missing or outlier data effectively through SQL conditional logic, ensuring cleaner and more reliable visualizations.

- **Dynamic plot preparation**: Enhance plotting workflows with context-aware SQL techniques, allowing adaptable and insightful data visualizations.

Looking Ahead

The next chapter, "Optimizing Your Script with Indexes and Views," explores techniques for improving the performance and efficiency of SQL queries. This includes understanding how indexes can speed up data retrieval by optimizing how the database accesses and organizes data, and the role that views play in simplifying complex queries by creating reusable, virtual tables that streamline workflows. By combining indexes and views, you can enhance query execution time while maintaining clarity and maintainability.

Test Your Skills

John is a data analyst at an international university tasked with analyzing students' overall performance across various courses. The university administration wants answers to questions, such as:

1. How do students' grades compare across different study programs, categorized into performance levels (e.g., High: 85+, Medium: 70–84, Low: <70)?

2. What percentage of students in each study program excel (grade $\geq$ 90) or struggle (grade $\leq$ 60) in their courses?

3. Which age groups tend to perform better on different courses?

4. How does student performance in core courses (e.g., Data Structures for CS, Economics for Business) vary across semesters?

5. What proportion of students improved their grades in the same course across semesters?

John realizes that static reports are insufficient. He needs dynamic, context-sensitive data visualizations that can adapt based on specific analysis criteria. John decides to use SQL for data preparation before the visualizations process. Use the data in Tables 8-16, 8-17, and 8-18 to answer these questions.

Table 8-16. *The Students Data Table*

student_id	student_name	Age	study_program
1	Alice Johnson	18	Computer Science
2	Bob Smith	22	Business Administration
3	Charlie Davis	19	Engineering
4	Diana King	21	Computer Science
5	Ethan White	20	Engineering
6	Fiona Brown	23	Business Administration
7	George Hall	18	Engineering
8	Hannah Scott	22	Computer Science
9	Ian Mitchell	19	Business Administration
10	Julia Lopez	21	Engineering
11	Kevin Turner	20	Computer Science
12	Laura Hughes	23	Business Administration
13	Mason Cox	19	Engineering
14	Nina Patel	22	Business Administration
15	Oliver Stone	18	Computer Science

Table 8-17. *The Courses Data Table*

course_id	course_name	study_program
1	Data Structures	Computer Science
2	Marketing 101	Business Administration
3	Physics	Engineering
4	Databases	Computer Science
5	Economics	Business Administration
6	Thermodynamics	Engineering
7	Machine Learning	Computer Science

Table 8-18. *The Grades Table*

grade_id	student_id	course_id	grade	semester
1	14	1	77	Spring 2024
2	4	6	83	Fall 2023
3	12	4	93	Spring 2024
4	12	6	55	Fall 2023
5	2	2	96	Fall 2023
6	4	6	88	Spring 2024
7	7	3	66	Spring 2024
8	6	1	89	Spring 2024
9	8	2	81	Fall 2023
10	6	2	62	Spring 2024
11	1	6	92	Spring 2024
12	11	7	52	Spring 2024
13	15	4	74	Spring 2024
14	7	4	89	Fall 2023
15	7	7	92	Spring 2024

CHAPTER 9

Optimizing Your Script with Indexes and Views

When working with large data volumes, optimizing SQL queries can significantly improve performance. This type of optimization can be achieved through indexes and views, which are both powerful tools. Indexes speed up data retrieval by allowing the database engine to locate rows more efficiently, reducing the need to scan entire tables. Views, on the other hand, simplify complex queries by storing reusable SQL logic, enhancing readability and maintainability. Additionally, materialized views reduce the need to recalculate data by storing precomputed results. Together, indexes and views help minimize query execution time, lower system resource consumption, and enhance the overall responsiveness of applications, making them vital for maintaining efficient and scalable database systems. This chapter examines how to utilize these capabilities in order to optimize the data analysis process.

Introduction

Indexes are database structures that improve data retrieval speed. They work like a book index, allowing the database to quickly locate rows without scanning the entire table. Indexes are a critical tool for optimizing SQL queries, and they offer several key benefits that make them essential for data analysts and database administrators. Faster query execution is one of the primary reasons for using indexes. Indexes significantly speed up operations like `SELECT`, `WHERE`, `JOIN`, and `ORDER BY` by allowing the database to quickly locate the relevant rows without scanning the entire table. This is particularly helpful for complex queries or those involving large datasets, as it reduces the time needed to retrieve results. Another advantage is efficient data retrieval. Instead of performing a full table scan, which can be resource-intensive and slow, the database can utilize the index to directly access the required rows, making queries more efficient. This efficiency

© Hamed Tabrizchi 2025
H. Tabrizchi, *Narrative SQL*, https://doi.org/10.1007/979-8-8688-1560-7_9

is especially valuable when working with large tables containing millions of rows. The performance difference between using an index and scanning the entire table can be substantial. By using indexes, data analysts can ensure queries run smoothly and quickly, even when dealing with massive amounts of data.

Views are virtual tables that display SQL query results. While it does not store data itself, it simplifies complex queries and abstracts the underlying data structures. Views are an essential tool in SQL that simplify and optimize data analysis workflows. One of the primary advantages of views is their ability to simplify complex queries. By breaking intricate queries into reusable components, views make it easier to write, read, and maintain SQL code. This saves time and reduces errors. Additionally, views provide data abstraction, allowing analysts to hide the complexity of the underlying database schema from end users. Through this abstraction, users can interact with the data without knowing the database structure. Another key benefit is security. Views can restrict access to specific columns or rows, ensuring sensitive information is only accessible to authorized users. Finally, views promote consistency by providing a standardized interface for frequently used queries. This consistency ensures that all users work with the same data definitions, reducing discrepancies and improving analysis reliability.

Understanding Indexes

An *index* in SQL is a data structure that improves a table's retrieval speed. It works like an index in a book—it helps you quickly find specific information without scanning the entire content. SQL indexes offer performance benefits, including faster search queries, improved sorting and filtering, and efficient joins between tables. However, they also consume additional disk space and can slow down INSERT, UPDATE, and DELETE operations due to the need to update the index.

In data analysis, while indexes improve query performance, they also have trade-offs that can impact data storage and write operations. They can affect disk space and slow down INSERT, UPDATE, and DELETE operations, which are used relatively less often in the data analysis process than SELECT is.

Indexes consume additional disk space as they store separate data structures referencing table rows. Especially for large datasets and multiple indexes, this can significantly increase storage costs and slow down database backups. Furthermore, indexes impact INSERT, UPDATE, and DELETE operations. Each time a row is modified, all associated indexes must also be updated, adding overhead and slowing down write operations. For example, inserting a large number of rows requires updating all indexes, significantly increasing insertion time. Therefore, a balance must be struck. In data analysis, where reads are frequent but writes are less so, indexes are highly beneficial. However, for tables with frequent writes, excessive indexing can degrade performance. Optimizations like partial indexes, which index only specific data subsets, as well as temporarily disabling indexes during bulk imports can mitigate these performance drawbacks.

Note   There are some database systems that handle indexing in a different way. For example, clustered indexes in SQL Server and index-organized tables in Oracle store the actual table data within the index itself. In these cases, the distinction between index and table data blurs, and storage or performance implications may differ slightly.

Basic Syntax for Creating an Index

The basic syntax for creating an index is provided here:

```
CREATE INDEX index_name ON table_name (column_name);
```

Here, `CREATE INDEX` is the command to create an index. `Index_name` is the name of the index. It is best to use a descriptive name. `ON table_name` specifies the table where the index is created, and `column_name` is the column or columns the index is based on.

Types of Indexes in PostgreSQL

Index types in PostgreSQL are divided into seven types—B-tree index, unique index, hash index, generalized inverted index, generalized search tree (GiST) index, partial index, and composite index. Each is discussed in more detail in this section.

B-tree Index

The *B-tree index* is the most common and default index type in PostgreSQL. It is particularly effective for equality comparisons using the = operator, as well as range queries involving operators like <, >, <=, and >=. Additionally, B-tree indexes significantly speed up sorting operations performed with the ORDER BY clause. By organizing data in a balanced tree structure, they allow quick traversal from root to leaf nodes, ensuring consistent performance even as the dataset grows. This makes B-tree indexes ideal for large tables that require frequent searches and ordered data retrieval.

```
CREATE INDEX idx_salary ON employees (salary);
```

Unique Index

A *unique index* ensures that all values in a specific column or combination of columns are distinct, preventing duplicate entries in the table. PostgreSQL automatically creates unique indexes when defining columns with PRIMARY KEY or UNIQUE constraints. For example, a primary key column uses an index to maintain data integrity by ensuring each value is unique and non-NULL. Similarly, a unique constraint enforces distinct values while allowing NULLs. Unique indexes are essential for maintaining data accuracy, supporting efficient lookups, and optimizing query performance. Particularly when validating user input or maintaining referential integrity.

```
CREATE UNIQUE INDEX idx_unique_name ON employees (name);
```

Hash Index

A *hash index* in PostgreSQL is optimized for simple equality comparisons using the = operator. Unlike B-tree indexes, which support both equality and range queries, hash indexes are specifically designed for exact matches, making them faster for such queries. This efficiency stems from hashing algorithms that map data values to fixed locations, enabling quick lookups. However, hash indexes are not suitable for range queries or ordering because they do not maintain data order. While they offer faster performance for exact matches, they require more careful use since they do not support advanced query types like <, >, or ORDER BY.

```
CREATE INDEX idx_hash_department ON employees USING HASH (department);
```

Generalized Inverted Index (GIN)

A *generalized inverted index* (GIN) is designed for indexing complex data structures such as arrays, JSON fields, and full-text search in PostgreSQL. Unlike traditional B-tree indexes, GIN efficiently handles multiple values stored within a single column by mapping each distinct element to the corresponding row entries.

```
CREATE INDEX idx_gin_name ON employees USING GIN (to_
tsvector('english', name));
```

This index enables fast lookup of words within the name column, significantly improving search performance in large datasets.

Generalized Search Tree (GiST) Index

A *generalized search tree* (GiST) index is a flexible, balanced-tree indexing mechanism used for complex data types such as geometric shapes, network addresses, and full-text search. Unlike B-trees, GiST indexes allow custom search strategies, making them essential for spatial queries and full-text indexing.

```
CREATE INDEX idx_gist_salary ON employees USING GiST (salary);
```

This allows PostgreSQL to quickly locate employees within specific salary bands, improving performance for range-based queries. Additionally, GiST indexes support nearest-neighbor searches, making them valuable in geospatial applications where distance-based queries are common.

Partial Index

A *partial index is* an optimized index that stores only a subset of table rows, reducing storage overhead and improving query performance. PostgreSQL avoids maintaining unnecessary index entries by indexing only records that meet a predefined condition, leading to faster lookups for queries that frequently filter on the indexed condition.

```
CREATE INDEX idx_high_salary ON employees (salary) WHERE salary > 50000;
```

This index is useful for queries frequently retrieving employees earning more than $50,000, making them significantly faster than scanning the entire table. Partial indexes are particularly effective in large datasets where a full index would be wasteful and unnecessary.

Composite Index

A *composite index* indexes multiple columns together, making it ideal for queries filtering or sorting by multiple attributes. Instead of creating separate indexes for each column, PostgreSQL uses a composite index to speed up queries that involve a combination of indexed columns in their filtering conditions.

```
CREATE INDEX idx_name_department ON employees (name, department);
```

This index improves performance when searching for employees by name within specific departments, as it allows PostgreSQL to efficiently navigate the data based on both columns simultaneously. Composite indexes are especially useful in multi-column filtering, sorting, and JOIN operations.

Note   For data analysis tasks, the most applicable PostgreSQL indexes are B-tree, GIN, and GiST due to their efficiency in handling large datasets, complex queries, and specific data types.

1. **B-tree index:** Ideal for equality comparisons (=), range queries (<, >, <=, >=), and sorting (ORDER BY). Its balanced structure ensures consistent performance, making it essential for filtering and aggregating large datasets.

2. **GIN index:** Efficient for indexing composite data types such as arrays, JSON, and full-text search. It's particularly useful for advanced data analysis tasks involving semi-structured data and quick lookups within nested structures.

3. **GiST index:** Best for multidimensional data and spatial analysis. It supports geometric data types, text search, and similarity searches, making it valuable for geographic information systems (GIS) and similarity-based data analysis.

4. **BRIN index:** Optimized for very large datasets, especially those with natural ordering, such as time-series data. By storing minimal information for each data block, BRIN reduces storage space and accelerates queries that scan large, sequential ranges.

Other index types, like Hash, are more specialized and less commonly used in general data-analysis scenarios. Unique indexes are essential for data integrity but do not directly impact analytical performance.

Dropping an Index

If you no longer need an index, you can remove it using the DROP INDEX command:

```
DROP INDEX idx_department;
```

Checking Index Usage with EXPLAIN

When you query PostgreSQL, you can use the EXPLAIN keyword to see if an index is being used:

```
EXPLAIN SELECT * FROM employees WHERE department = 'IT';
```

Look for the Index Scan in the output, which indicates that the index is being used. If you see Seq Scan, PostgreSQL scans the entire table (which is slower).

When to Use and When to Avoid Indexes

Database indexes are powerful tools for optimizing query performance, but their use should be strategic. Indexes are highly beneficial when dealing with large, frequently querying tables, especially when searching, filtering, sorting, or joining on specific columns. Indexing can, however, be detrimental if it is excessive. Because indexes consume additional disk space and slow down INSERT, UPDATE, and DELETE operations, it's crucial to avoid over-indexing. A balanced approach is key, carefully selecting which columns to index based on query patterns and data modification frequency. An overview explanation of when to use and when to avoid indexes is provided in Table 9-1.

Table 9-1. *When to Use and When to Avoid Indexes*

Use Indexes When	Avoid Excessive Indexing Because
The table is large and frequently queried	Indexes consume disk space (excessive indexing increases storage costs)
Specific columns are often searched, filtered, or sorted	Indexes slow down INSERT, UPDATE, and DELETE operations
Queries involve joins between large tables	Maintaining multiple indexes adds overhead during data modifications
Queries frequently use WHERE, ORDER BY, or GROUP BY	Indexes need to be updated each time data is modified, slowing down writes
Full-text search is performed on large text fields (It is recommended to use GIN)	Too many indexes can confuse the query planner, leading to suboptimal execution
The table has foreign keys or needs to enforce uniqueness	Indexes may not provide benefits for small tables or low-selectivity columns
Queries filter by JSON fields or array elements (It is recommended to use GIN or GiST)	Indexes can degrade performance if queries rarely use the indexed columns
Time-series or sequential data requires efficient range queries (It is recommended to use BRIN)	Partial indexes or composite indexes might be more efficient in some scenarios

The Role of Indexes in Data Analysis Tasks

An index acts as a guide within a database for data retrieval, helping to speed up data analysis. This allows the database to skip full table scans and quickly locate specific data points. Especially with large datasets, exhaustive searches would be prohibitively slow. By optimizing filtering and sorting operations, indexes enable analysts to rapidly narrow down data based on specific criteria or arrange it meaningfully. This is essential for tasks like identifying trends, outliers, or patterns in the data.

Indexes also boost data warehousing and analytical performance by facilitating joins between multiple tables. The index facilitates faster joins for a holistic view of data from disparate sources. As well as supporting range queries, they also accelerate the calculation of sums, averages, and counts, which are vital to summarizing and understanding data distributions. Indexes enable data analysts to perform complex investigations, extract valuable insights, and make data-driven decisions more quickly. Although there is a tradeoff in terms of storage space and writing performance, the benefits outweigh the disadvantages for workloads that involve a substantial amount of reading and writing.

Using EXPLAIN to Review Query Execution

EXPLAIN helps analyze how PostgreSQL plans to execute a query, which is essential when optimizing analytical queries on large datasets.

```
EXPLAIN query_statement;
```

Here, `query_statement` is the SQL query to analyze and provides insights into how the database engine intends to execute the SQL statement. In a nutshell, PostgreSQL analyzes and parses the SQL statement provided by `EXPLAIN query_statement`. Instead of actually executing the query and returning the results, it generates an execution plan. This plan outlines the steps to execute the query. As output, this plan shows the estimated cost and sequence of operations.

Note   In relation to scan types, EXPLAIN can reveal whether the database performs a sequential scan, reading every row of a table, or if it uses an index for efficient retrieval via an index scan, index only scan using only index data, or bitmap scan using bitmap indexes. When queries involve JOINs, EXPLAIN details the JOIN method employed, such as `NestedLoop`, iterative matching, `MERGE JOIN`, merging sorted rows, or `Hash Join`, hash table probing. Furthermore, the output clearly displays the filtering conditions applied to the data, so you can understand how the database narrows down the results.

Using EXPLAIN ANALYZE for Performance Measurement

EXPLAIN ANALYZE executes the query and provides actual runtime statistics. It is best to use this for testing performance improvements after adding indexes.

```
EXPLAIN ANALYZE query_statement;
```

Here, query_statement is the SQL query to analyze with real-time metrics. EXPLAIN ANALYZE is a tool for database performance analysis in PostgreSQL. It goes beyond the estimated costs provided by a simple EXPLAIN by actually executing the query and collecting real-time metrics. Table 9-2 indicates the key differences of EXPLAIN and EXPLAIN ANALYZE.

Table 9-2. *Key Differences Between EXPLAIN and EXPLAIN ANALYZE*

Feature	EXPLAIN	EXPLAIN ANALYZE
Actual execution	Generates an execution plan (no execution)	Executes the query; gathers runtime stats
Real-time metrics	Provides estimated costs	Provides actual execution times, row counts

In addition to estimating costs and providing a plan structure, EXPLAIN ANALYZE provides runtime metrics that enhance the analysis. In PostgreSQL, the planner's cost estimates are expressed as two numbers: a startup cost and a total cost. These are unitless values based on estimated I/O, CPU usage, and row counts, and are used to compare alternative execution plans rather than reflect actual time. EXPLAIN ANALYZE reveals the actual execution time for each node, measured in milliseconds, which directly highlights performance bottlenecks. The actual row counts allow for a precise comparison of the planner's estimates, revealing their accuracy. In nested loop joins, loop counts specify the frequency of inner loop executions. Additionally, other runtime statistics may also be included, depending on the specific query and PostgreSQL version. This can provide an overview of the query's real-world execution.

The First Story: Golf Performance Data Analysis

Martina is a data analyst at a golf club, responsible for evaluating player performance and optimizing tournament operations. She manages a database that tracks players' scores, course details, and tournament results. The data tables shown in Tables 9-3, 9-4, and 9-5 are growing as more tournaments take place, and she wants to ensure that the queries run efficiently. With indexes and performance analysis tools like EXPLAIN ANALYZE, Martina aims to speed up data retrieval and facilitate better decision-making.

Table 9-3. *The Players Table*

player_id	Name	Age	Nationality
1	Tiger Woods	48	USA
2	Rory McIlroy	34	Northern Ireland
3	Jon Rahm	29	Spain
4	Brooks Koepka	33	USA
5	Dustin Johnson	40	USA
6	Collin Morikawa	27	USA
7	Justin Thomas	31	USA
8	Phil Mickelson	53	USA
9	Jordan Spieth	31	USA
10	Viktor Hovland	26	Norway

Table 9-4. *The Tournaments Table*

tournament_id	Name	Location	Year
1	Masters Tournament	Augusta National Golf Club	2023
2	US Open	Pebble Beach	2023
3	The Open Championship	St Andrews	2023
4	PGA Championship	Oak Hill	2023
5	Ryder Cup	Marco Simone	2023

Table 9-5. *The Scores Table*

score_id	player_id	tournament_id	round_number	Strokes
1	1	1	1	70
2	1	1	2	72
3	2	1	1	68
4	2	1	2	70
5	3	2	1	71
6	3	2	2	69
7	4	2	1	70
8	4	2	2	68
9	5	3	1	73
10	5	3	2	72

Martina creates indexes on columns frequently used for filtering and joining to improve query performance.

```
CREATE INDEX idx_player_id ON scores(player_id);
CREATE INDEX idx_tournament_id ON scores(tournament_id);
CREATE INDEX idx_name ON players(name);
```

The columns Martina chose for indexing are frequently used for filtering, joining, and sorting. These are the primary use cases where indexes improve data analysis efficiency. For instance, an index on the scores(player_id) column can speed up queries that filter or join by player ID, which is common when retrieving a specific player's scores. Since player_id is a foreign key in the Scores table, indexing it improves join performance with the Players table. To filter scores by tournament, the scores(tournament_id) column is essential for analyzing player performance at a specific event. Indexing tournament_id speeds up queries that group, filter, or join scores from the Tournaments table. For the players(name)column, an index on name supports direct lookups of players by name, which is common when analyzing an individual player's performance. Although names are not unique, indexing this column still enhances search efficiency.

These indexes were selected to balance performance improvements for frequent data analysis tasks while minimizing overhead associated with excessive indexing.

To get answers to the following questions, Martina intends to use this data and SQL queries:

- Who scored below 70 strokes in any round of the Masters Tournament?

- Calculate the average strokes for each player in the U.S. Open.

- Find the top three players with the lowest combined scores in the Open Championship.

- Retrieve all scores of Tiger Woods for performance analysis.

Martina wrote the following query to find players who scored below 70 strokes in any round of the Masters Tournament. Martina uses EXPLAIN to show whether indexes are used for JOINs and filtering. An index scan should replace seq scan for better performance. This query benefits from an index on player_id and tournament_id since it filters by both.

```
EXPLAIN SELECT p.name, s.strokes
FROM scores s
JOIN players p ON s.player_id = p.player_id
JOIN tournaments t ON s.tournament_id = t.tournament_id
WHERE t.name = 'Masters Tournament' AND s.strokes < 70;
```

The query retrieves player names and strokes for scores under 70 in the Masters Tournament. It joins three tables: Scores, Players, and Tournaments. The JOIN condition links scores.player_id with players.player_id and scores.tournament_id with tournaments.tournament_id. The WHERE clause filters the results to include only scores less than 70 and tournaments named Masters Tournament.

Due to the EXPLAIN command, executing this query will show these results:

```
                             QUERY PLAN
------------------------------------------------------------------------
-----------
 Nested Loop  (cost=13.92..47.53 rows=4 width=122)
   -> Hash Join  (cost=13.78..46.54 rows=4 width=8)
         Hash Cond: (s.tournament_id = t.tournament_id)
```

```
->  Seq Scan on scores s  (cost=0.00..31.25 rows=567 width=12)
       Filter: (strokes < 70)
->  Hash  (cost=13.75..13.75 rows=2 width=4)
       ->  Seq Scan on tournaments t  (cost=0.00..13.75 rows=2
           width=4)
              Filter: ((name)::text = 'Masters Tournament'::text)
->  Index Scan using players_pkey on players p  (cost=0.15..0.25 rows=1
    width=122)
       Index Cond: (player_id = s.player_id)
(10 rows)
```

Here, the EXPLAIN output shows how PostgreSQL executes the query. It starts with a sequential scan on the scores table, filtering rows where strokes < 70, costing 31.25. Then, a hash scan on the tournaments table filters for "Masters Tournament," costing 13.75. A hash join connects scores and tournaments, costing 46.54. Finally, a nested loop joins this result with the Players table using an index scan on the primary key, which costs 0.25. Indexing on players.player_id improves performance, but sequential scans on scores and tournaments suggest more optimization strategies.

In order to calculate the average strokes for each player in the U.S. Open, Martina wrote the following query:

```
EXPLAIN ANALYZE SELECT p.name, AVG(s.strokes) AS avg_strokes
FROM scores s
JOIN players p ON s.player_id = p.player_id
JOIN tournaments t ON s.tournament_id = t.tournament_id
WHERE t.name = 'US Open'
GROUP BY p.name;
```

The query calculates the average strokes of players who participated in the U.S. Open tournament. It joins the scores, players, and tournaments tables using player_id and tournament_id. The WHERE clause filters scores by the U.S. Open tournament, and the GROUP BY clause groups results by player names. The AVG() function computes the average strokes for each player. This query is common in data analysis and business intelligence to assess player performance in specific tournaments. Proper indexing improves joining efficiency and reduces query execution time, especially when working with large datasets containing multiple tournaments and players.

QUERY PLAN

```
-----------------------------------------------------------------------------
-----------------------------------------------------------
 GroupAggregate  (cost=47.62..47.84 rows=11 width=150) (actual
time=0.137..0.156 rows=2 loops=1)
   Group Key: p.name
   -> Sort  (cost=47.62..47.65 rows=11 width=122) (actual
      time=0.124..0.138 rows=4 loops=1)
      Sort Key: p.name
      Sort Method: quicksort  Memory: 25kB
      -> Nested Loop  (cost=13.92..47.43 rows=11 width=122) (actual
         time=0.057..0.095 rows=4 loops=1)
         -> Hash Join  (cost=13.78..45.30 rows=11 width=8) (actual
            time=0.041..0.061 rows=4 loops=1)
            Hash Cond: (s.tournament_id = t.tournament_id)
            -> Seq Scan on scores s  (cost=0.00..27.00 rows=1700
               width=12) (actual time=0.004..0.013 rows=10
               loops=1)
            -> Hash  (cost=13.75..13.75 rows=2 width=4) (actual
               time=0.013..0.015 rows=1 loops=1)
               Buckets: 1024  Batches: 1  Memory Usage: 9kB
               -> Seq Scan on tournaments t  (cost=0.00..13.75
                  rows=2 width=4) (actual time=0.004..0.007
                  rows=1 loops=1)
                  Filter: ((name)::text = 'US Open'::text)
                  Rows Removed by Filter: 4
         -> Index Scan using players_pkey on players
            p  (cost=0.15..0.19 rows=1 width=122) (actual
            time=0.004..0.004 rows=1 loops=4)
            Index Cond: (player_id = s.player_id)
 Planning Time: 0.179 ms
 Execution Time: 0.227 ms
(18 rows)
```

The EXPLAIN ANALYZE output shows the query's execution plan with actual execution times. The process begins with a sequential scan on the Scores table, retrieving ten rows in 0.013 ms. A sequential scan on the Tournaments table filters for "US Open," retrieving one row in 0.007 ms. A hash join connects scores and tournaments in 0.061 ms, followed by an index scan using players_pkey for each matched row, which is efficient, cost: 0.19. The results are sorted using quicksort, Memory: 25kB, and a GroupAggregate calculates the average strokes. The total execution time is 0.227 ms, which is considered efficient given the query's simplicity, the small number of rows processed, and minimal I/O—especially when compared to more complex queries or poorly optimized joins.

To find the top three players with the lowest combined scores in the Open Championship, the following query calculates the total strokes of players who participated in "The Open Championship" tournament. It joins the Scores, Players, and Tournaments tables using player_id and tournament_id. The WHERE clause filters scores for "The Open Championship," and the GROUP BY clause groups results by player names. The SUM() function calculates each player's total strokes. The ORDER BY total_strokes ASC arranges results in ascending order, and LIMIT 3 returns the three players with the fewest total strokes.

```
EXPLAIN SELECT p.name, SUM(s.strokes) AS total_strokes
FROM scores s
JOIN players p ON s.player_id = p.player_id
JOIN tournaments t ON s.tournament_id = t.tournament_id
WHERE t.name = 'The Open Championship'
GROUP BY p.name
ORDER BY total_strokes ASC
LIMIT 3;
```

The EXPLAIN output reveals the query's execution plan. The process starts with a sequential scan on the Scores table, retrieving 1,700 rows, followed by a sequential scan on the Tournaments table, filtering for "The Open Championship." A hash join merges these tables. Next, an index scan uses players_pkey to retrieve player names efficiently. The results are sorted twice, first by player names using Sort, then by total strokes using another Sort with cost=47.96. The Limit step returns the top three players. As shown here, the query plan indicates efficient performance due to proper indexing and sorting techniques:

QUERY PLAN

```
---------------------------------------------------------------------
---------------------------------------
 Limit  (cost=47.96..47.96 rows=3 width=126)
   -> Sort  (cost=47.96..47.98 rows=11 width=126)
         Sort Key: (sum(s.strokes))
         -> GroupAggregate  (cost=47.62..47.81 rows=11 width=126)
               Group Key: p.name
               -> Sort  (cost=47.62..47.65 rows=11 width=122)
                     Sort Key: p.name
                     -> Nested Loop  (cost=13.92..47.43 rows=11 width=122)
                           -> Hash Join  (cost=13.78..45.30 rows=11
                              width=8)
                              Hash Cond: (s.tournament_id =
                              t.tournament_id)
                              -> Seq Scan on scores
                                 s  (cost=0.00..27.00 rows=1700
                                 width=12)
                              -> Hash  (cost=13.75..13.75 rows=2
                                 width=4)
                                    -> Seq Scan on tournaments
                                       t  (cost=0.00..13.75 rows=2
                                       width=4)
                                        Filter: ((name)::text = 'The
                                        Open Championship'::text)
                           -> Index Scan using players_pkey on players
                              p  (cost=0.15..0.19 rows=1 width=122)
                              Index Cond: (player_id = s.player_id)
(16 rows)
```

To retrieve all scores of Tiger Woods for performance analysis, the following query retrieves tournament names, round numbers, and strokes for Tiger Woods. It joins the Scores, Players, and Tournaments tables using player_id and tournament_id. The WHERE clause filters the Players table to include only rows where the name matches

"Tiger Woods." The query outputs relevant scores and round numbers from the Scores table along with tournament names from the Tournaments table. This query is useful for analyzing a player's performance across different tournaments, helping coaches, analysts, and fans evaluate consistency and identify trends in performance across multiple rounds.

```
EXPLAIN ANALYZE SELECT t.name, s.round_number, s.strokes
FROM scores s
JOIN players p ON s.player_id = p.player_id
JOIN tournaments t ON s.tournament_id = t.tournament_id
WHERE p.name = 'Tiger Woods';
```

The EXPLAIN ANALYZE output shows the query's execution steps. First, a sequential scan of the Scores table retrieves 1,700 rows. A sequential scan of the Players table finds the player "Tiger Woods" by filtering nine rows. A hash join merges both tables, matching player_id values. Next, an index scan using tournaments_pkey retrieves tournament names based on tournament_id, executed twice. The total query execution took 0.107 ms, indicating fast performance due to indexing and the limited number of matching rows.

```
                              QUERY PLAN
-----------------------------------------------------------------------------
----------------------------------------------------------
 Nested Loop  (cost=14.55..47.85 rows=10 width=126) (actual
time=0.041..0.082 rows=2 loops=1)
   -> Hash Join  (cost=14.40..45.92 rows=10 width=12) (actual
       time=0.027..0.059 rows=2 loops=1)
       Hash Cond: (s.player_id = p.player_id)
       -> Seq Scan on scores s  (cost=0.00..27.00 rows=1700 width=16)
           (actual time=0.005..0.014 rows=10 loops=1)
       -> Hash  (cost=14.38..14.38 rows=2 width=4) (actual
           time=0.016..0.018 rows=1 loops=1)
           Buckets: 1024  Batches: 1  Memory Usage: 9kB
           -> Seq Scan on players p  (cost=0.00..14.38 rows=2 width=4)
               (actual time=0.004..0.007 rows=1 loops=1)
               Filter: ((name)::text = 'Tiger Woods'::text)
               Rows Removed by Filter: 9
```

```
 -> Index Scan using tournaments_pkey on tournaments t  (cost=0.15..0.19
    rows=1 width=122) (actual time=0.006..0.006 rows=1 loops=2)
       Index Cond: (tournament_id = s.tournament_id)
Planning Time: 0.101 ms
Execution Time: 0.107 ms
(13 rows)
```

Data analysis tasks like filtering by player names, joining large tables, and aggregating scores are significantly faster with indexes. Using EXPLAIN and EXPLAIN ANALYZE helps confirm that PostgreSQL utilizes indexes for better performance. This optimization is crucial for business intelligence, enabling faster decision-making based on player performance and tournament outcomes.

Martina wants to delete an index at this stage for a number of reasons. First, if the index is no longer used by queries, it unnecessarily consumes storage and slows down write operations like INSERT, UPDATE, and DELETE, in future queries. Second, an index can become inefficient if data distribution changes over time, requiring re-creation. Lastly, removing redundant indexes improves overall database performance, especially when maintaining multiple indexes for overlapping purposes. To delete an index in PostgreSQL, use the DROP INDEX statement. When Martina wants to delete the index called players_pkey, she can do so as follows:

```
DROP INDEX players_pkey;
```

In this story, the emphasis was on how to optimize query execution, and how to evaluate this optimization. As mentioned, indexes have a positive effect when dealing with large data tables and massive amounts of data, but such tables do not easily fit within the pages of this book! Therefore, the purpose of this story is limited to proving insight into the process of optimizing using INDEX and evaluating its effectiveness using EXPLAIN or EXPLAIN ANALYZE.

Understanding SQL Views

In PostgreSQL, a *view* is essentially a virtual table derived from the result of a stored query. It doesn't store data itself; instead, it provides a customized, simplified view of the underlying data from one or more tables.

SQL views in PostgreSQL offer a powerful way to streamline database interactions. The concept of a view can be described as a virtual table created as a result of a stored query. A key advantage of views is that they simplify complex queries by encapsulating them, allowing them to be reused; instead of repeatedly writing elaborate queries, you can use views as standard tables. The use of views enhances the security of data by restricting access to specific columns or rows, allowing users to see only the information they require. In addition, views provide data abstraction, protecting applications from schema changes even when table structure alterations are made. Views also improve readability by simplifying complex queries into smaller, more manageable components.

Basic Syntax for SQL Views

The basic syntax of an SQL view is provided here:

```
CREATE VIEW view_name AS
SELECT column1, column2, ...
FROM table1
WHERE condition;
```

Here, `CREATE VIEW` creates a virtual table, called a view. It stores an SQL query that dynamically generates data when called, without storing the data itself. The syntax begins with `CREATE VIEW` followed by the desired view name. The `AS` keyword introduces the query that defines the view. The `SELECT` statement specifies the columns to include, the source table using `FROM`, and optional filtering conditions using `WHERE`.

Types of Views in PostgreSQL

There are two ways to implement views in PostgreSQL—through the use of standard views and materialized views. Standard views, created with `CREATE VIEW`, act as virtual tables, presenting the results of stored queries without physically storing the data. This makes them perpetually up-to-date, reflecting any changes to the underlying tables. In contrast, materialized views, created with `CREATE MATERIALIZED VIEW`, store the query's result as a physical table. This creates a snapshot of the data at the time of creation or refresh.

This storage mechanism significantly enhances performance for frequently accessed, complex queries, but requires manual or scheduled data refreshes. A trade-off between real-time data accuracy and query performance optimization determines the choice between standard and materialized views (see Table 9-6).

Table 9-6. *Key Differences Between Standard Views and Materialized Views*

Feature	Standard Views	Materialized Views
Data storage	Virtual; no physical data stored	Physical table; data is stored
Data currency	Always up-to-date	Snapshot; requires manual refresh
Performance	Query performance depends on underlying tables	Improved performance for complex queries
Use case	Simplification, security, abstraction	Performance optimization, reporting

The Role of Views in Data Analysis Tasks

Views can play a crucial role in SQL data analysis, serving as a layer of abstraction that significantly enhances the analytical process' efficiency. They simplify complex data models by presenting focused, pre-aggregated, and filtered representations, reducing query complexity and increasing readability. By encapsulating intricate logic, views act as reusable analysis building blocks, ensuring consistency and accelerating analysis through pre-computed summaries and segmented data subsets. In addition to speeding up data processing, this improves accuracy and productivity, allowing analysts to concentrate on extracting insights rather than retrieving data. In addition, views serve to enhance data security by restricting access to sensitive information, making them essential for streamlined and secure analytics.

The Second Story: Car Race Data Analysis

Nathalie, a data analyst for Speed Track Racing, needs to analyze race performance data from recent competitions. As shown in Tables 9-7, 9-8, and 9-9, the database contains three main tables: Drivers, Races, and Results. By creating SQL views and using the data in these tables, Nathalie wants to optimize repetitive queries and improve analysis efficiency.

Table 9-7. *The Drivers Table*

driver_id	Name	Team
1	Lewis Hamilton	Mercedes
2	Max Verstappen	Red Bull
3	Charles Leclerc	Ferrari
4	Sergio Perez	Red Bull
5	Carlos Sainz	Ferrari
6	George Russell	Mercedes
7	Lando Norris	McLaren
8	Fernando Alonso	Aston Martin
9	Pierre Gasly	Alpine
10	Esteban Ocon	Alpine

Table 9-8. *The Races Table*

race_id	Location	Date
101	Silverstone	2023-07-09
102	Monza	2023-09-03
103	Spa-Francorch.	2023-07-30
104	Suzuka	2023-09-24
105	Monaco	2023-05-28
106	Austin	2023-10-22
107	Singapore	2023-09-17
108	Interlagos	2023-11-05
109	Budapest	2023-07-23
110	Melbourne	2023-04-02

Table 9-9. *The Results Table*

result_id	driver_id	race_id	position	lap_time_sec
1	1	101	1	90.23
2	2	101	2	90.45
3	3	101	3	91.12
4	4	102	1	87.32
5	5	102	2	87.78
6	6	103	1	89.01
7	7	103	2	89.23
8	8	104	1	88.45
9	9	104	3	89.67
10	10	105	1	92.89

Nathalie aims to answer the following questions based on race performance data collected from recent races.

- Who are the top performers who finished first in each race?

- What is the average lap time for each driver across all races?

- How many first-place finishes does each team have?

- Who were the winners of each race, along with race location and date?

- Who recorded the fastest lap in each race?

Nathalie wrote the following query to identify drivers who consistently performed well across tracks. This task can be made easier with views, which store logic for joining the Drivers, Races, and Results tables. Instead of rewriting complex JOINs every time, analysts can query the top_performers view directly. This improves readability, reduces errors, and ensures consistency.

```
CREATE VIEW top_performers AS
SELECT d.name, r.location, res.position
FROM results res
```

```
JOIN drivers d ON res.driver_id = d.driver_id
JOIN races r ON res.race_id = r.race_id
WHERE res.position = 1;
```

This SQL query creates a view called top_performers, which displays the race winners. This query selects driver names, race locations, and finishing positions, and joins these three tables: Results, Drivers, and Races. It then connects these tables using driver_id and race_id as keys, and filters to only include first-place finishes (position = 1). Based on the filter, the view will display the name of each winning driver, the location where they won, and their position. Table 9-10 illustrates what is happening in this query as a result of the SELECT statement.

Table 9-10. *The Race Winners*

Name	Location	Position
Lewis Hamilton	Silverstone	1
Sergio Perez	Monza	1
George Russell	Spa-Francorch.	1
Fernando Alonso	Suzuka	1
Esteban Ocon	Monaco	1

The following query was written by Nathalie to find the average lap time for each driver across all races, and to answer the question of driver consistency. This simplifies reporting, especially if this metric is needed in multiple dashboards or reports. Views also ensure that there is only one source of truth, keeping calculations consistent across all analyses.

```
CREATE VIEW avg_lap_time AS
SELECT d.name, AVG(res.lap_time_sec) AS avg_time
FROM results res
JOIN drivers d ON res.driver_id = d.driver_id
GROUP BY d.name;
```

This SQL query creates a view called avg_lap_time, which calculates the average lap time for each driver. It selects driver names from the Drivers table and computes the average lap times, in seconds, from the Results table. The query joins these tables using driver_id as the common key, allowing it to match lap times with the corresponding drivers. Results are grouped by driver name, meaning the view displays each driver's name alongside their average lap time across all recorded races. Using this view, it is possible to quickly compare the overall performance of drivers when it comes to their speed. The driver names are illustrated in Table 9-11, along with the average lap time, in seconds, calculated for each driver.

Table 9-11. *The Average Lap Time for Each Driver*

Name	avg_time
Sergio Perez	87.31999969
Lando Norris	89.2300034
Pierre Gasly	89.66999817
Lewis Hamilton	90.2300034
Fernando Alonso	88.44999695
Max Verstappen	90.44999695
Esteban Ocon	92.88999939
George Russell	89.01000214
Charles Leclerc	91.12000275
Carlos Sainz	87.77999878

In order to determine how many first-place finishes each team has, Nathalie wrote the following query for a question that evaluates team performance by summing the victories of each team's drivers.

```
CREATE VIEW team_wins AS
SELECT d.team, COUNT(*) AS wins
FROM results res
JOIN drivers d ON res.driver_id = d.driver_id
WHERE res.position = 1
GROUP BY d.team;
```

This SQL query creates a view named team_wins, which tracks the number of race victories for each racing team. It joins the Results and Drivers tables to connect race performance with team information. The query counts only first-place finishes, where position is equal to 1, and groups the results by team name. This means the view will display each team's total wins across all races. By using COUNT(*), it counts the number of races where a team's driver finished first, providing a clear overview of team performance and success in racing competitions.

Here, a view named team_wins simplifies data analysis by creating a predefined, easy-to-access summary of racing performance. Views allow quick retrieval of complex information—in this case, team victories—without repeatedly writing complicated SQL joins. Views also provide a convenient, performance-efficient way to track and analyze team success across multiple races. Table 9-12 illustrates the SELECT output of the previous query, which is the number of race victories for each racing team.

Table 9-12. *The Number of Race Victories for Each Racing Team*

Team	Wins
Alpine	1
Aston Martin	1
Mercedes	2
Red Bull	1

The following query aims to find race winners in each race along with the race location and date. This question highlights race winners while connecting them to the location and date for easy event tracking.

```
CREATE VIEW race_winners AS
SELECT d.name, r.location, r.date
FROM results res
JOIN drivers d ON res.driver_id = d.driver_id
JOIN races r ON res.race_id = r.race_id
WHERE res.position = 1;
```

This SQL query creates a view named `race_winners`. It selects the winner's name, race location, and date by joining the `Results`, `Drivers`, and `Races` tables. The `JOIN` clauses link the driver and race information to race results. The `WHERE` clause filters for only results where `position` equals 1, identifying the race winners. As a result of creating a view, future queries will be simplified. Rather than repeatedly writing this complex `JOIN`, analysts can query `race_winners` for a concise list of race champions. This improves readability and efficiency. Table 9-13 illustrates the winner's name, race location, and date.

Table 9-13. The Winner's Name, Race Location, and Date

Name	Location	Date
Lewis Hamilton	Silverstone	2023-07-09
Sergio Perez	Monza	2023-09-03
George Russell	Spa-Francorch.	2023-07-30
Fernando Alonso	Suzuka	2023-09-24
Esteban Ocon	Monaco	2023-05-28

This query identifies who recorded the fastest lap in each race and highlights standout lap performances regardless of race position.

```
CREATE VIEW fastest_laps AS
SELECT d.name, r.location, MIN(res.lap_time_sec) AS fastest_time
FROM results res
JOIN drivers d ON res.driver_id = d.driver_id
JOIN races r ON res.race_id = r.race_id
GROUP BY r.location, d.name;
```

This query creates a view called `fastest_laps`. It aims to identify the fastest lap time achieved by each driver at each race location. The query joins the `Results`, `Drivers`, and `Races` tables to link driver names and race locations with lap times. The `MIN(res.lap_time_sec)` function determines the shortest lap time for each driver at each location. The `GROUP BY r.location, d.name` clause organizes the results, ensuring that the minimum lap time is calculated separately for each driver at every race location. This view simplifies accessing and analyzing fastest lap data.

Note   In PostgreSQL, the MIN() function is an aggregate function that returns the smallest value within a specified set of values. It's commonly used to find the minimum value of a column in a table. As soon as MIN() is applied, it analyzes the selected column and returns the lowest value. For example, it can be applied to identify the earliest date, the lowest price, or the fastest lap time. To find the minimum value within specific groups, it's often combined with GROUP BY.

Table 9-14 provides the fastest lap time achieved by each driver at each race location.

Table 9-14. *The Fastest Lap Time Achieved by Each Driver at Each Race Location*

Name	Location	fastest_time
Carlos Sainz	Monza	87.78
Sergio Perez	Monza	87.32
Pierre Gasly	Suzuka	89.67
Charles Leclerc	Silverstone	91.12
Max Verstappen	Silverstone	90.45
Fernando Alonso	Suzuka	88.45
Lando Norris	Spa-Francorch.	89.23
George Russell	Spa-Francorch.	89.01
Esteban Ocon	Monaco	92.89
Lewis Hamilton	Silverstone	90.23

Views simplify data access by storing predefined queries, reducing code repetition, and improving readability. By optimizing complex joins and aggregations, views make analysis faster and more efficient. With these views, Nathalie has quick access to essential race performance insights without having to repeatedly write complex SQL queries.

Managing Views

As mentioned, the `CREATE VIEW` statement in PostgreSQL defines a virtual table based on the result set of an SQL statement. The syntax for `CREATE VIEW` is straightforward, allowing for the creation of a named query that can be used as a table, thus simplifying complex queries and enhancing data security.

Updating and Modifying Views (ALTER VIEW)

While PostgreSQL doesn't have a direct `ALTER VIEW` statement to modify the query definition of a view, it is possible to modify a view with `CREATE OR REPLACE VIEW`. This command allows you to redefine the view's query without dropping and re-creating it, preserving any dependent objects. The syntax is similar to `CREATE VIEW`, but using `CREATE OR REPLACE VIEW` ensures that if the view already exists, its definition is replaced. Maintaining consistency and avoiding disruptions is crucial.

PostgreSQL provides an `ALTER VIEW` statement for modifying view properties, such as renaming the view or changing its column names and default values. The basic syntax includes the following.

Renaming a view:

```
ALTER VIEW view_name RENAME TO new_view_name;
```

Changing column names within a view:

```
ALTER VIEW view_name RENAME COLUMN old_column_name TO new_column_name;
```

Setting or removing default values for view columns:

```
ALTER VIEW view_name ALTER COLUMN column_name SET DEFAULT default_value;
ALTER VIEW view_name ALTER COLUMN column_name DROP DEFAULT;
```

As a result, these commands can maintain view consistency without requiring complete redefinition. This makes updates more efficient and minimizes interruptions to dependent objects.

Dropping Views (DROP VIEW)

The DROP VIEW statement is used to remove a view from the database. The syntax is simple:

```
DROP VIEW view_name;
```

This command permanently deletes the view definition. Whenever a view is dropped, any queries or applications that rely on it will fail. To enhance flexibility, PostgreSQL provides a number of additional options that can be used.

Avoiding errors if the view does not exist:

```
DROP VIEW IF EXISTS view_name;
```

This prevents errors in scripts and automated processes where the existence of a view cannot be guaranteed.

Dropping multiple views at once:

```
DROP VIEW view_name1, view_name2, ...;
```

Using RESTRICT to prevent accidental deletions:

```
DROP VIEW view_name RESTRICT;
```

This ensures that the view will be dropped only if no other objects are dependent on it. In addition to managing database integrity, these options allow for controlled cleanup of unused views, preventing unintended disruptions.

The Role of ALTER VIEW and DROP VIEW in Data Analysis

In data analysis, views play a critical role in structuring and simplifying complex queries, making data retrieval more efficient. The ability to update a view using CREATE OR REPLACE VIEW ensures that analysts can modify data representations without disrupting workflows. This is particularly beneficial in cases where data models evolve, and existing views need to be adjusted without affecting dependent reports or applications. On the other hand, DROP VIEW is essential for maintaining a clean and optimized database by removing outdated or unnecessary views. The IF EXISTS option, however, is particularly useful in dynamic and automated environments, where analysts must exercise caution when dropping views in order to prevent breaking dependent queries.

The Role of Views in Optimizing SQL Queries

When views are used correctly, they reduce redundancy by encapsulating complex logic, thus avoiding the repetition of lengthy SQL code across multiple scripts. In this way, maintainability is enhanced, and inconsistencies are reduced. A view effectively balances storage, speed, and query complexity. As a result of abstracting complex queries, views offer simplified interfaces that reduce the need for intricate SQL in everyday analysis. Especially materialized views provide speed advantages by precompiling and storing results, although at the expense of storage. While standard views do not physically store data, they streamline query complexity, resulting in a more manageable and cleaner script. As a result of the strategic implementation of views, data access is optimized, ensuring that queries are both efficient and readable, thus optimizing overall performance and maintainability.

Using Both Views and Indexes in PostgreSQL

Views and indexes are complementary but serve different purposes in SQL. Views are virtual tables that store a query's result definition, making complex queries more readable and reusable. They simplify data access, maintain consistency, and improve security by restricting access to specific columns. Indexes speed up data retrieval by allowing the database to locate rows more efficiently instead of scanning entire tables. They are essential for performance optimization, especially when dealing with large datasets.

Since views do not store data, they do not automatically benefit from indexes unless they are materialized or used with indexed queries. The purpose of this section is to explore how to use both effectively.

The Third Story: Online Retail Data Analyst

Kris is a data analyst at an online retail company. The company's sales team frequently asks Kris for reports on product performance. However, querying the raw sales table every time is slow, especially as the database grows. To solve this, Kris decides to use views for structured queries and indexes for better performance.

Note   There is no doubt that, in real-world scenarios, datasets are significantly larger and more complex than the sample presented in Table 9-15. Due to space constraints in presenting stories in the book, the table is limited to ten rows. As long as the same methods are applied to large datasets, this limitation does not impact query effectiveness.

Table 9-15. *The Sales Table*

sale_id	product_id	amount	sale_date
1	101	120.5	2024-03-01
2	102	85.75	2024-03-02
3	101	99.99	2024-03-02
4	103	150	2024-03-03
5	101	110	2024-03-04
6	104	200	2024-03-05
7	102	80.5	2024-03-06
8	103	175.25	2024-03-07
9	101	125	2024-03-08
10	105	300	2024-03-09

Kris examines the Sales table, which tracks customer purchases including the sale_id, product_id, amount (sales value), and sale_date.

Kris notices that most reports filter data by sale_date, meaning that the database has to scan the entire table to find relevant records. To improve this, Kris creates an index on the sale_date column:

```
CREATE INDEX idx_sale_date ON sales(sale_date);
```

Now, whenever a query filters by sale_date, PostgreSQL uses the index, significantly speeding up performance.

As a next step, Kris uses a view to calculate total sales per product instead of manually aggregating sales data:

```
CREATE VIEW sales_summary AS
SELECT product_id, SUM(amount) AS total_sales
FROM sales
GROUP BY product_id;
```

With this view, she can quickly retrieve total sales without rewriting complex queries.

Now, in order to eliminate the need for each query to recalculate the data, Kris creates a materialized view that stores the computed results:

```
CREATE MATERIALIZED VIEW sales_summary_mat AS
SELECT product_id, SUM(amount) AS total_sales
FROM sales
GROUP BY product_id;
```

To speed up lookups, Kris adds an index on product_id:

```
CREATE INDEX idx_sales_summary ON sales_summary_mat(product_id);
```

Instead of reprocessing the entire dataset whenever the sales team requests a report, Kris queries the materialized view:

```
SELECT * FROM sales_summary_mat;
```

Table 9-16 illustrates the total sales of each product separately.

Table 9-16. *The Total Sales of Each Product*

product_id	total_sales
104	200
105	300
103	325.25
101	455.49
102	166.25

Since new sales are recorded daily, Kris refreshes the materialized view periodically:

```
REFRESH MATERIALIZED VIEW sales_summary_mat;
```

As mentioned earlier, materialized views store query results physically instead of recalculating them every time. The REFRESH MATERIALIZED VIEW sales_summary_ mat; command must be used to keep the data current, as a materialized view does not automatically update when the underlying table changes. Without running this command, the materialized view continues to display outdated data. This is even if updated sales records are added, updated, or deleted from the Sales table. This can lead to discrepancies in reports and inaccurate business decisions. By refreshing the materialized view, the latest sales data can be reflected in the summary, allowing the sales team to work with accurate and current information. While refreshing materialized views can be resource-intensive, it is essential for maintaining data integrity, especially in environments where reports and analytics rely on fresh data.

The sales team now gets their reports instantly, and Kris no longer has to run slow, repetitive queries manually

The use of both views and indexes in SQL has several advantages. The optimization of performance is one of the key benefits of views, since they simplify queries by structuring complex data retrieval, while indexes enhance query execution by reducing the need for full table scans. In addition to improving security and access control, views can restrict access to specific columns or join multiple tables to present only relevant data, while indexes ensure that access to this data is as efficient as possible. Furthermore, materialized views with indexing offer significant performance improvements by storing query results physically, enabling indexes to be applied directly to the results. This can reduce computation time and increase query efficiency, particularly for data that is frequently accessed.

Summary

This chapter highlighted the power of SQL's index and view in optimizing data analysis. In this chapter, you learned how to improve query performance and data management. SQL index enables faster data retrieval by reducing the need for full table scans. Views allow structured and simplified query execution by encapsulating complex logic. Also, using both of them can significantly improve efficiency, maintainability, and security in database operations.

Key Points

- Indexes enhance query performance by reducing the need for full table scans, enabling faster data retrieval from large datasets.

- Views simplify complex queries by encapsulating logic into reusable structures, making data analysis more efficient and maintainable.

- Materialized views store query results physically, allowing indexing optimizations for faster reporting and analytical processing.

- The combination of indexes and views enhances efficiency, ensuring that frequently accessed data is both structured and retrievable rapidly.

Key Takeaways

- **Optimized query performance**: Indexes reduce query execution time by minimizing full table scans, improving data retrieval efficiency.

- **Simplified data access**: Views transform complex queries into reusable structures, enhancing readability and maintainability.

- **Faster analytical processing**: Materialized views store precomputed results, allowing indexed optimization to be performed for faster insight.

- **Efficient data management**: Combining views with indexes ensures structured access to frequently queried data with minimal overhead.

Looking Ahead

The next chapter, "Analytics Alchemy: Turning Data into Gold, explores powerful techniques that transform raw data into meaningful insights. It investigates advanced analytical functions that enable deeper data exploration and window functions for data comparison, which allow you to analyze trends and relationships across rows, as well as strategies for turning raw data into compelling stories through insightful structured reporting.

Test Your Skills

Angela is a database administrator at a large e-commerce company responsible for optimizing data retrieval and reporting. The company's business intelligence team needs fast and efficient access to sales data to answer key performance questions, such as:

- What are the top-selling products in each category over the past six months?

- How does revenue growth compare across different regions and time periods?

- Which customers make frequent high-value purchases, and what products do they prefer?

- How do seasonal trends impact product demand across different categories?

- What percentage of orders are returned, and how does this affect overall revenue?

Angela realizes that executing complex queries repeatedly on massive sales datasets slows down reporting and analytics. To address this, she decides to use indexes to speed up searches on frequently queried columns, such as `order_date`, `customer_id`, and `product_category`. Additionally, she creates views to simplify data retrieval for analysts by predefining commonly used queries, such as total revenue per region or top-selling products. See Tables 9-17, 9-18, and 9-19.

Table 9-17. *The Orders Table*

order_id	customer_id	product_id	order_date	quantity	total_price	Region
1001	501	2001	2024-01-05	2	50	East
1002	502	2003	2024-02-12	1	120	West
1003	503	2002	2024-02-15	5	200	South
1004	504	2001	2024-03-01	3	75	East
1005	501	2004	2024-03-10	1	300	North
1006	505	2002	2024-03-18	4	160	South
1007	506	2005	2024-04-02	2	90	West
1008	507	2003	2024-04-08	3	360	North
1009	508	2004	2024-04-15	2	600	East
1010	509	2001	2024-04-20	1	25	South

Table 9-18. *The Products Table*

product_id	product_name	category	unit_price	stock_quantity
2001	Notebook	Stationery	25	500
2002	Laptop	Electronics	40	100
2003	Smartphone	Electronics	120	50
2004	Office Chair	Furniture	300	30
2005	Headphones	Electronics	45	200

Table 9-19. *The Customers Table*

customer_id	Name	Age	Region	total_spent
501	Alice	32	East	375
502	Bob	45	West	120
503	Charlie	28	South	200
504	David	40	East	75
505	Eva	35	South	160
506	Frank	50	West	90
507	Grace	27	North	360
508	Henry	41	East	600
509	Irene	30	South	25

CHAPTER 10

Analytics Alchemy: Turning Data into Gold

Turning data into gold with SQL requires mastering advanced analytical functions that help you extract deeper insights from raw data and transform them into compelling narratives. PostgreSQL provides powerful functions that can be used to transform raw data into a compelling story. Additionally, this chapter explains how you can transform raw data into a compelling story, and how recursive queries can be used to structure query logic effectively.

Functions

Functions can be categorized based on their functionality and purpose. PostgreSQL provides a wide range of functions, and while aggregate, statistical and mathematical, and window functions are common categories, there are also other types, including ranking functions, string functions, date and time functions, JSON functions, control functions and system functions. SQL functions help you perform operations that would otherwise require several queries and round trips within the database. By using functions, it is possible to reuse your database since other applications can interact directly with your stored procedures without the need for middle-tier or duplicate code. This is valuable for most data analysis tasks.

Aggregate Functions

Aggregate functions operate on multiple rows and return a single summarized value. They are commonly used with GROUP BY to perform calculations across subsets of data. For instance, PostgreSQL aggregate functions are used to find the total revenue per product category. Aggregate functions operate on multiple rows and return a

© Hamed Tabrizchi 2025
H. Tabrizchi, *Narrative SQL*, https://doi.org/10.1007/979-8-8688-1560-7_10

single summarized value for each group of data. Common aggregate functions include SUM(column_name), which calculates the total sum of a column, and AVG(column_name), which determines the average value. Additionally, COUNT(column_name) is used to count the number of rows, while MIN(column_name) and MAX(column_name) help identify the smallest and largest values in a dataset. In a nutshell, the key feature of aggregate functions is that they collapse multiple rows into a single result per group. This makes them ideal for summarizing large datasets efficiently. Table 10-1 summarizes most of the well-known aggregate functions in PostgreSQL.

Table 10-1. *The Commonly Used Aggregate Functions in PostgreSQL*

Function	Description	Example
AVG(column_name)	Returns the average (arithmetic mean) of the input values.	SELECT AVG(salary) FROM employees;
COUNT(column_name)	Returns the number of input rows for which the value of expression is not null.	SELECT COUNT(department) FROM employees;
COUNT(*)	Returns the number of input rows.	SELECT COUNT(*) FROM employees;
MAX(column_name)	Returns the maximum value of the input values.	SELECT MAX(salary) FROM employees;
MIN(column_name)	Returns the minimum value of the input values.	SELECT MIN(salary) FROM employees;
SUM(column_name)	Returns the sum of the input values.	SELECT SUM(salary) FROM employees;
BOOL_AND(column_name)	Returns TRUE if all input values are true, otherwise FALSE.	SELECT BOOL_AND(active) FROM users;
BOOL_OR(column_name)	Returns TRUE if any input value is true, otherwise FALSE.	SELECT BOOL_OR(active) FROM users;
EVERY(column_name)	Equivalent to BOOL_AND.	SELECT EVERY(active) FROM users;

(continued)

Table 10-1. (*continued*)

Function	Description	Example
`MODE(column_name)`	Returns the most frequent input value (or any if multiple are equally frequent).	`SELECT MODE() WITHIN GROUP (ORDER BY department) FROM employees;`
`PERCENTILE_CONT(fraction)`	Returns a percentile, interpolating between adjacent input items if needed.	`SELECT PERCENTILE_CONT(0.5) WITHIN GROUP (ORDER BY salary) FROM employees;`
`PERCENTILE_DISC(fraction)`	Returns a percentile; the result is an actual input element.	`SELECT PERCENTILE_DISC(0.9) WITHIN GROUP (ORDER BY salary) FROM employees;`

Statistical and Mathematical Functions

Statistical and mathematical functions perform advanced numerical computations and are often used for predictive modeling and trend analysis. A practical use case for statistical and mathematical functions in PostgreSQL is calculating the correlation between advertising spend and sales.

Note   Correlation in statistics measures the strength and direction of a relationship between two variables. It quantifies how changes in one variable relate to changes in another. The correlation coefficient (r) ranges from -1 to 1:

- $r = 1 \rightarrow$ Perfect positive correlation (both increase together).

- $r = -1 \rightarrow$ Perfect negative correlation (one increases, the other decreases).

- $r = 0 \rightarrow$ No correlation (no relationship).

The most common correlation method is Pearson's correlation, which assumes a linear relationship. Other types include Spearman's, for ranked data, and Kendall's, for ordinal data.

As a result of statistical and mathematical functions, it is possible to go beyond simple aggregates in order to provide statistical insights and numerical modeling. For instance, REGR_SLOPE(y, x) and REGR_INTERCEPT(y, x) perform linear regression analysis, helping to identify trends between two variables.

Note   Linear regression analysis is a statistical method used to model the relationship between two variables by fitting a straight line. It predicts how a dependent variable (Y) changes based on an independent variable (X). The equation is:

$Y=mX+b$

Where m is the slope (rate of change), and b is the intercept (Y's value when X = 0). Linear regression analysis helps us in trend analysis, forecasting, and decision-making. Assumptions include linearity, independence, homoscedasticity, and normality of residuals for accurate predictions.

To measure data dispersion, functions like STDDEV(column_name) and VARIANCE(column_name) quantify variability in datasets.

Note   Data dispersion, or variability, refers to how spread out or scattered the values in a dataset are. It helps to understand the degree of variation within the data. Common measures of dispersion include range, variance, and standard deviation. The range is the difference between the maximum and minimum values. Variance is the average squared deviation of each data point from the mean, showing how data points differ from the average. Standard deviation is the square root of variance, providing a measure of how much individual data points deviate from the mean.

Higher dispersion indicates more variability, while lower dispersion means the data points are closer to the mean.

Mathematical transformations, such as POWER(x, y), LOG(x), and SQRT(x) further assist in adjusting and normalizing data for analysis. The key advantage of these functions is their ability to extract patterns, relationships, and insights that are critical for data-driven decision-making. Table 10-2 summarizes most of the well-known statistical and mathematical functions in PostgreSQL.

Table 10-2. *Commonly Used Statistical and Mathematical Functions*

Category	Function	Description	Example
Mathematical	ABS(x)	Returns the absolute value of x.	SELECT ABS(-5);
Mathematical	CBRT(x)	Returns the cube root of x.	SELECT CBRT(27);
Mathematical	CEIL(x) or CEILING(x)	Returns the smallest integer greater than or equal to x.	SELECT CEIL(4.2);
Mathematical	EXP(x)	Returns e raised to the power of x.	SELECT EXP(1);
Mathematical	FLOOR(x)	Returns the largest integer less than or equal to x.	SELECT FLOOR(4.8);
Mathematical	LN(x) or LOG(x)	Returns the natural logarithm of x.	SELECT LN(10);
Mathematical	LOG(b, x)	Returns the logarithm of x to base b.	SELECT LOG(10, 100);
Mathematical	MOD(y, x)	Returns the remainder of y divided by x.	SELECT MOD(11, 3);
Mathematical	PI()	Returns the value of pi.	SELECT PI();
Mathematical	POWER(x, y) or POW(x, y)	Returns x raised to the power of y.	SELECT POWER(2, 3);
Mathematical	ROUND(x)	Rounds x to the nearest integer.	SELECT ROUND(4.5);
Mathematical	ROUND(x, s)	Rounds x to s decimal places.	SELECT ROUND(4.567, 2);
Mathematical	SIGN(x)	Returns the sign of x (-1, 0, or 1).	SELECT SIGN(-10);

(continued)

Table 10-2. (*continued*)

Category	Function	Description	Example
Mathematical	SQRT(x)	Returns the square root of x.	SELECT SQRT(16);
Mathematical	TRUNC(x)	Truncates x toward 0.	SELECT TRUNC(4.8);
Mathematical	TRUNC(x, s)	Truncates x to s decimal places.	SELECT TRUNC(4.567, 2);
Mathematical	RANDOM()	Returns a pseudo-random number between 0 and 1.	SELECT RANDOM();
Statistical	AVG(expression)	Returns the average (arithmetic mean) of the input values.	SELECT AVG(salary) FROM employees;
Statistical	CORR(y, x)	Returns the correlation coefficient.	SELECT CORR(sales, advertising) FROM marketing;
Statistical	COVAR_POP(y, x)	Returns the population covariance.	SELECT COVAR_POP(sales, advertising) FROM marketing;
Statistical	COVAR_SAMP(y, x)	Returns the sample covariance.	SELECT COVAR_SAMP(sales, advertising) FROM marketing;
Statistical	STDDEV_POP(expression)	Returns the population standard deviation.	SELECT STDDEV_POP(salary) FROM employees;
Statistical	STDDEV_SAMP(expression)	Returns the sample standard deviation.	SELECT STDDEV_SAMP(salary) FROM employees;

(*continued*)

Table 10-2. (*continued*)

Category	Function	Description	Example
Statistical	VARIANCE (expression)	Returns the sample variance.	SELECT VARIANCE(salary) FROM employees;
Statistical	VAR_ POP(expression)	Returns the population variance.	SELECT VAR_ POP(salary) FROM employees;
Statistical	VAR_ SAMP(expression)	Returns the sample variance.	SELECT VAR_ SAMP(salary) FROM employees;
Statistical	REGR_AVGX(y, x)	Returns the average of the independent variable (x).	SELECT REGR_AVGX(y, x) FROM table;
Statistical	REGR_AVGY(y, x)	Returns the average of the dependent variable (y).	SELECT REGR_AVGY(y, x) FROM table;
Statistical	REGR_INTERCEPT (y, x)	Returns the y-intercept of the least-squares regression line.	SELECT REGR_ INTERCEPT(y, x) FROM table;
Statistical	REGR_R2(y, x)	Returns the square of the correlation coefficient.	SELECT REGR_R2(y, x) FROM table;
Statistical	REGR_SLOPE(y, x)	Returns the slope of the least-squares regression line.	SELECT REGR_SLOPE(y, x) FROM table;
Statistical	REGR_COUNT(y, x)	Returns the number of input rows where both expressions are not null.	SELECT REGR_COUNT(y, x) FROM table;

Window Functions

Unlike aggregate functions, window functions do not collapse rows but instead return a computed value for each row while maintaining access to the full dataset. The key advantage of window functions is that they allow row-wise comparisons and trend analysis, unlike aggregate functions. This allows analysts to maintain granular data while uncovering insights. Table 10-3 summarizes most of the well-known window functions in PostgreSQL.

Table 10-3. *The Commonly Used Window Functions*

Function	Description	Example
ROW_NUMBER()	Assigns a unique sequential integer to each row within the partition.	SELECT ROW_NUMBER() OVER (PARTITION BY department ORDER BY salary DESC) FROM employees;
CUME_DIST()	Calculates the cumulative distribution of a value within the partition.	SELECT CUME_DIST() OVER (PARTITION BY department ORDER BY salary DESC) FROM employees;
NTILE(n)	Divides the partition into n approximately equal groups and assigns a group number to each row.	SELECT NTILE(4) OVER (PARTITION BY department ORDER BY salary DESC) FROM employees;
LAG(expression [, offset [, default]])	Accesses data from a previous row in the partition.	SELECT LAG(salary, 1, 0) OVER (PARTITION BY department ORDER BY hire_date) FROM employees;
LEAD(expression [, offset [, default]])	Accesses data from a subsequent row in the partition.	SELECT LEAD(salary, 1, 0) OVER (PARTITION BY department ORDER BY hire_date) FROM employees;

(continued)

Table 10-3. (*continued*)

Function	Description	Example
FIRST_ VALUE(expression)	Returns the value of the expression from the first row in the partition.	SELECT FIRST_VALUE(salary) OVER (PARTITION BY department ORDER BY hire_date) FROM employees;
LAST_ VALUE(expression)	Returns the value of the expression from the last row in the partition.	SELECT LAST_VALUE(salary) OVER (PARTITION BY department ORDER BY hire_date RANGE BETWEEN UNBOUNDED PRECEDING AND UNBOUNDED FOLLOWING) FROM employees;
NTH_ VALUE(expression, n)	Returns the value of the expression from the nth row in the partition.	SELECT NTH_VALUE(salary, 3) OVER (PARTITION BY department ORDER BY hire_date) FROM employees;
COALESCE(value1, value2, ...)	Returns the first non-NULL value in the list.	SELECT COALESCE(NULL, 'default', 'other');` (returns 'default')
NULLIF(value1, value2)	Returns NULL if value1 equals value2; otherwise returns value1.	SELECT NULLIF(10, 10);` (returns NULL) `SELECT NULLIF(10, 5);` (returns 10)
GREATEST(value1, value2, ...)	Returns the largest value from the list.	SELECT GREATEST(10, 5, 15); (returns 15)
LEAST(value1, value2, ...)	Returns the smallest value from the list.	SELECT LEAST(10, 5, 15); (returns 5)
ARRAY[value1, value2, ...]	Constructs an array from the given values.	SELECT ARRAY[1, 2, 3]; (returns {1, 2, 3})
GENERATE_ SERIES(start, stop [, step])	Generates a series of values.	SELECT GENERATE_SERIES(1, 5); (returns 1, 2, 3, 4, 5)
GENERATE_ SUBSCRIPTS(array, dimension)	Generates the set of valid subscripts for the specified dimension of the given array.	SELECT GENERATE_ SUBSCRIPTS(ARRAY['a','b','c'], 1); (returns 1, 2, 3)

As shown in Table 10-3, a special type of window functions are value functions. Using value functions, you can manipulate individual data values, including type conversion, NULL handling, and data formatting. These functions operate on single values within a column or expression, enabling precise data manipulation.

Ranking Functions

Ranking functions are essential for assigning order to rows within a query's result. They provide a way to rank data based on specified criteria, facilitating analysis and reporting. RANK() assigns a rank to each row, skipping ranks when equal ranks occur. DENSE_ RANK() also assigns ranks, but doesn't skip ranks for ties, ensuring sequential ranking. ROW_NUMBER() assigns a unique sequential integer to each row, regardless of equality. These functions use the OVER() clause to define the result set ordering and partitioning. They are invaluable for tasks like identifying top performers, analyzing sales data, and generating leaderboards, providing clear and structured insights from ordered data. Table 10-4 summarizes most of the well-known ranking functions in PostgreSQL.

Table 10-4. *Commonly Used Ranking Functions*

Function	Description	Example
RANK()	Assigns a rank to each row within the partition, with gaps for tied values.	SELECT RANK() OVER (PARTITION BY department ORDER BY salary DESC) FROM employees;
DENSE_ RANK()	Assigns a rank to each row within the partition, without gaps for tied values.	SELECT DENSE_RANK() OVER (PARTITION BY department ORDER BY salary DESC) FROM employees;
ROW_ NUMBER()	Assigns a unique sequential integer to each row within the partition.	SELECT ROW_NUMBER() OVER (PARTITION BY department ORDER BY salary DESC) FROM employees;
NTILE(n)	Divides the partition into n approximately equal groups and assigns a group number to each row.	SELECT NTILE(4) OVER (PARTITION BY department ORDER BY salary DESC) FROM employees;
PERCENT_ RANK()	Calculates the relative rank of each row within the partition (0 to 1).	SELECT PERCENT_RANK() OVER (PARTITION BY department ORDER BY salary DESC) FROM employees;

String Functions

String functions are essential for text manipulation, enabling tasks like joining, extracting, and formatting text data. For instance, CONCAT() joins multiple strings into one, simplifying string composition. LOWER() and UPPER() convert strings to lowercase or uppercase, ensuring case consistency. TRIM() removes leading and trailing spaces, cleaning up text data. LENGTH() returns the character count of a string. REPLACE() substitutes specified substrings, which is useful for data correction. These functions are indispensable for data cleaning, formatting, and analysis, making text data more manageable and consistent within databases. Table 10-5 summarizes most of the well-known string functions in PostgreSQL.

Table 10-5. *Commonly Used String Functions*

Function	Description	Example
CONCAT(string1, string2, ...) or string1 ‖ string2 ‖ ...	Concatenates strings.	SELECT CONCAT('Post', 'greSQL'); (returns 'PostgreSQL')
LEFT(string, n)	Returns the first n characters of a string.	SELECT LEFT('PostgreSQL', 4); (returns 'Post')
RIGHT(string, n)	Returns the last n characters of a string.	SELECT RIGHT('PostgreSQL', 4);` (returns 'SQL')
TRIM([LEADING | TRAILING | BOTH] [characters] FROM string)	Removes leading, trailing, or both occurrences of characters from a string.	SELECT TRIM(' PostgreSQL '); (returns 'PostgreSQL')
LTRIM(string [, characters])	Removes leading characters from a string.	SELECT LTRIM(' PostgreSQL'); (returns 'PostgreSQL')
RTRIM(string [, characters])	Removes trailing characters from a string.	SELECT RTRIM('PostgreSQL '); (returns 'PostgreSQL')
UPPER(string)	Converts a string to uppercase.	SELECT UPPER('PostgreSQL'); (returns 'POSTGRESQL')

(continued)

Table 10-5. (*continued*)

Function	Description	Example
LOWER(string)	Converts a string to lowercase.	SELECT LOWER('PostgreSQL'); (returns 'postgresql')
LENGTH(string)	Returns the length of a string.	SELECT LENGTH('PostgreSQL'); (returns 10)
POSITION(substring IN string)	Returns the position of the first occurrence of a substring in a string. When you use the POSITION function, the position returned starts at 1, not 0. This is unlike many programming languages, like C and Python, where indexing starts at 0.	`SELECT POSITION('gres' IN 'PostgreSQL'); (returns 6)
STRPOS(string, substring)	Returns the position of the first occurrence of a substring in a string.	SELECT STRPOS('PostgreSQL', 'gres'); (returns 6)
REPLACE(string, from, to)	Replaces all occurrences of a substring with another substring.	SELECT REPLACE('PostgreSQL', 'Post', 'New'); (returns 'NewgreSQL')
INITCAP(string)	Converts the first letter of each word to uppercase and the rest to lowercase.	SELECT INITCAP('postgreSQL database'); (returns 'PostgreSQL Database')
REPEAT(string, count)	Repeats a string a specified number of times.	SELECT REPEAT('Post', 3); (returns 'PostPostPost')
SPLIT_PART(string, delimiter, field)	Splits a string into parts using a delimiter and returns the specified field.	SELECT SPLIT_PART('PostgreSQL,Database', ',', 2); (returns 'Database')

(*continued*)

Table 10-5. (*continued*)

Function	Description	Example
`TO_CHAR(value, format)`	Converts a value to a string using a format.	`SELECT TO_CHAR(NOW(), 'YYYY-MM-DD');` (returns `'2023-10-27'` for example)
`MD5(string)`	Calculates the MD5 hash of a string.	`SELECT MD5('PostgreSQL');`
`SHA256(string)`	Calculates the SHA256 hash of a string.	`SELECT SHA256('PostgreSQL');`
`TRANSLATE(string, from, to)`	Replaces characters in a string with other characters.	`SELECT TRANSLATE('12345', '13', 'ax');` (returns `'a2x45'`)

Date and Time Functions

Date and time functions are crucial for handling temporal data, enabling precise timestamp manipulation. `NOW()` returns the current date and time, capturing the present moment. `AGE()` calculates the interval between two timestamps, useful for determining how time much has elapsed. `EXTRACT()` retrieves specific date or time components, like a year, month, or hour, for detailed analysis. `DATE_TRUNC()` truncates timestamps to a specified precision, such as a day or minute, for data aggregation. `TO_CHAR()` formats timestamps into custom string representations, facilitating report generation. These functions are essential for tasks involving time-series analysis, and data reporting, ensuring accurate and efficient temporal data management. Table 10-6 summarizes most of the well-known date and time functions in PostgreSQL.

Table 10-6. *The Commonly Used Date and Time Functions*

Function	Description	Example
NOW() or CURRENT_TIMESTAMP	Returns the current date and time with the time zone.	SELECT NOW();
CURRENT_DATE	Returns the current date.	SELECT CURRENT_DATE;
CURRENT_TIME	Returns the current time with the time zone.	SELECT CURRENT_TIME;
DATE(timestamp)	Extracts the date part from a timestamp.	SELECT DATE('2023-10-27 10:30:00'); (returns '2023-10-27')
EXTRACT(field FROM timestamp)	Extracts a specific field (e.g., year, month, day, hour, minute, second) from a timestamp.	SELECT EXTRACT(YEAR FROM '2023-10-27 10:30:00'); (returns 2023)
AGE(timestamp1, timestamp2) or AGE(timestamp)	Calculates the difference between two timestamps or the age of a timestamp relative to now.	SELECT AGE('2023-10-27', '2023-10-20'); (returns '7 days')
DATE_TRUNC(field, timestamp)	Truncates a timestamp to a specified field (e.g., day, month, year).	SELECT DATE_TRUNC('month', '2023-10-27 10:30:00'); (returns '2023-10-01 00:00:00')
TO_CHAR(timestamp, format)	Formats a timestamp as a string according to a format string.	SELECT TO_CHAR('2023-10-27 10:30:00', 'YYYY-MM-DD HH24:MI:SS'); (returns '2023-10-27 10:30:00')
TO_TIMESTAMP(string, format)	Converts a string to a timestamp according to a format string.	SELECT TO_TIMESTAMP('2023-10-27 10:30:00', 'YYYY-MM-DD HH24:MI:SS');

(*continued*)

Table 10-6. (*continued*)

Function	Description	Example
MAKE_DATE(year, month, day)	Constructs a date from year, month, and day.	SELECT MAKE_DATE(2023, 10, 27); (returns '2023-10-27')
MAKE_TIME(hour, minute, second)	Constructs a time from hour, minute, and second.	SELECT MAKE_TIME(10, 30, 0); (returns '10:30:00')
MAKE_TIMESTAMP(year, month, day, hour, minute, second)	Constructs a timestamp from year, month, day, hour, minute, and second.	SELECT MAKE_TIMESTAMP(2023, 10, 27, 10, 30, 0);
ISFINITE(timestamp)	Checks if a timestamp is finite (not infinity or -infinity).	SELECT ISFINITE('2023-10-27'); (returns true)
INTERVAL 'value unit'`	Constructs an interval value.	SELECT INTERVAL '1 day';
DATE_PART(text, timestamp)	Equivalent to extract, but returns a double precision number.	SELECT DATE_PART('day', timestamp '2023-10-27 12:00:00');
JUSTIFY_ DAYS(interval)	Adjusts an interval so that 30-day time spans are represented as months.	SELECT JUSTIFY_DAYS(INTERVAL '30 days');
JUSTIFY_ HOURS(interval)	Adjusts an interval so that 24-hour time spans are represented as days.	SELECT JUSTIFY_HOURS(INTERVAL '24 hours');
JUSTIFY_ INTERVAL(interval)	Adjusts an interval using both JUSTIFY_DAYS and JUSTIFY_HOURS.	SELECT JUSTIFY_ INTERVAL(INTERVAL '54 hours 31 days');

JSON Functions

JSON functions are designed to handle JSON data types, which are crucial for working with semi-structured data.

Note   JSON (JavaScript Object Notation) is a lightweight data-interchange format, ideal for representing structured data as key-value pairs. It's human-readable and easily parsed by machines. In SQL databases like PostgreSQL, JSON integration allows storing and querying semi-structured data alongside relational data. The ability to handle flexible data models, such as configuration settings or document-style data, that do not fit neatly into traditional tables, is crucial.

Using JSON functions in SQL, you can query and manipulate JSON data directly, bridging the gap between structured and semi-structured data. These functions empower analysts to query, modify, and aggregate JSON data directly within PostgreSQL, enhancing flexibility in data management. Table 10-7 summarizes most of the well-known JSON functions in PostgreSQL.

Table 10-7. *The Commonly Used JSON Functions*

Function	Description	Example
to_ json(anyelement)	Converts a PostgreSQL value to JSON.	SELECT to_json(ARRAY[1, 2, 3]); (returns "[1,2,3]")
to_ jsonb(anyelement)	Converts a PostgreSQL value to JSONB.	SELECT to_jsonb(ROW(1, 'foo')); (returns "{\"f1\":1,\"f2\":\"foo\"}")
json_build_ array(VARIADIC "any")	Builds a JSON array from a variadic parameter list.	SELECT json_build_array(1, 'two', null); (returns "[1,\"two\",null]")
json_build_ object(VARIADIC "any")	Builds a JSON object from a variadic parameter list.	SELECT json_build_object('foo', 1, 'bar', 'baz'); (returns "{\"foo\": 1,\"bar\":\"baz\"}")

Control Functions

Control functions enable procedural logic within database functions and triggers. IF-ELSE structures enable conditional execution, which enables branching logic based on data values. LOOP constructs provide iterative capabilities, facilitating repetitive tasks. EXCEPTION handling captures and manages errors, ensuring robust code execution. These control structures empower developers to create complex, dynamic database operations, such as data validation, conditional updates, and custom error handling. They are essential for building sophisticated database applications that require procedural logic beyond simple SQL queries. Table 10-8 summarizes most of the well-known control functions in PostgreSQL.

Table 10-8. *Commonly Used Control Functions*

Function	Description	Example
COALESCE(value1, value2, ...)	Returns the first non-NULL value in the list.	SELECT COALESCE(NULL, 'default', 'other'); (returns 'default')
NULLIF(value1, value2)	Returns NULL if value1 equals value2; otherwise returns value1.	SELECT NULLIF(10, 10); (returns NULL)
GREATEST(value1, value2, ...)`	Returns the largest value from the list.	SELECT GREATEST(10, 5, 15); (returns 15)
LEAST(value1, value2, ...)`	Returns the smallest value from the list.	SELECT LEAST(10, 5, 15); (returns 5)
CASE WHEN condition1 THEN result1 [WHEN condition2 THEN result2 ...] [ELSE resultN] END	Conditional expression, similar to if-then-else logic.	SELECT CASE WHEN age >= 18 THEN 'Adult' ELSE 'Minor' END FROM users;
ASSERT(condition, message)	Checks a condition and raises an error if it's false. (Primarily for debugging.)	SELECT ASSERT(age > 0, 'Age must be positive') FROM users;

System Functions

System functions provide access to database metadata and system-level operations. CURRENT_USER reveals the current user's name, aiding in security and auditing. VERSION() displays the PostgreSQL server's version, crucial for compatibility checks. pg_sleep() pauses execution for a specified duration, useful for testing and scheduling. These functions allow developers to interact with the database environment, retrieve system information, and control execution flow. Table 10-9 summarizes most of the well-known system functions in PostgreSQL.

Table 10-9. *Commonly Used System Functions*

Function	Description	Example
CURRENT_USER	Returns the current user name.	SELECT CURRENT_ USER;
SESSION_USER	Returns the session user name.	SELECT SESSION_ USER;
USER	Equivalent to CURRENT_USER.	SELECT USER;
CURRENT_DATABASE()	Returns the name of the current database.	SELECT CURRENT_ DATABASE();
CURRENT_SCHEMA	Returns the name of the current schema.	SELECT CURRENT_ SCHEMA;
CURRENT_ SCHEMAS(boolean)	Returns the names of schemas in the search path. If the Boolean parameter is true, implicitly included system schemas are included in the result.	SELECT CURRENT_ SCHEMAS(false);
VERSION()	Returns a string describing the PostgreSQL server version.	SELECT VERSION();

A summary of the differences between the types of functions is provided in Table 10-10.

Table 10-10. *Differences Between the Types of Functions*

Function Type	Purpose	Returns
Aggregate functions	Summarize multiple rows into one result.	Single value per group
Statistical and mathematical functions	Perform numerical or statistical calculations.	Single or multi-row values
Window functions	Compute values across rows without collapsing them.	One value per row
Value functions	Handle individual data values, often for type conversion or NULL handling.	Single value
Ranking functions	Rank rows within a result set based on specified criteria.	One value per row
String functions	Manipulate string data.	Single value
Date and time functions	Perform calculations or transformations on dates and times.	Single value or multi-row values
JSON functions	Manipulate or extract data from JSON/JSONB data types.	Single or multi-row values
Control functions	Control program flow in PostgreSQL.	Depends on the function (e.g., Boolean, Integer)
System functions	Interact with the database system or metadata.	Single value

Each function type serves a different analytical need; these are summarized and compared in Table 10-11.

Table 10-11. *Different Analytical Needs for Each Function*

Function Type	Description
Aggregate functions	Operate on multiple rows and return a single result.
Statistical and mathematical functions	For calculations like regression, dispersion, and other numerical analysis.
Window functions	Perform calculations across rows related to the current row, allowing comparisons.
Value functions	Return or manipulate values, often for type casting or handling NULL values.
Ranking functions	Rank rows based on specified ordering criteria.
String functions	For string manipulation tasks, such as concatenation and case conversion.
Date and time functions	Work with date and time types for calculations and transformations.
JSON functions	Manipulate and extract data from JSON/JSONB columns.
Control functions	Control program flow in PostgreSQL.
System functions	Deal with system-level information or database metadata.

Creating Your Own Functions in PostgreSQL

Creating your own functions in PostgreSQL is a powerful way to encapsulate reusable logic and extend your database's functionality. Here is the basic syntax for creating your own PostgreSQL functions:

```
CREATE FUNCTION function_name (parameters)
RETURNS return_type AS $$
DECLARE
    -- variable declarations (optional)
```

```
BEGIN
    -- function logic
    RETURN result;
END;
$$ LANGUAGE plpgsql;
```

Where CREATE FUNCTION function_name defines the function name, parameters are the input parameters the function takes (which is optional), and RETURNS return_type specifies the return type of the function (for instance, INTEGER, TEXT, or VOID for no return). DECLARE is optional and it declares any local variables used within the function. BEGIN and END mark the start and end of the function logic, and RETURN specifies the value the function returns.

For example, it might be useful to create a function that calculates the area of a rectangle. The following function takes two parameters, width and height, and returns the area as an INTEGER.

```
CREATE FUNCTION calculate_area(width INTEGER, height INTEGER)
RETURNS INTEGER AS $$
BEGIN
    RETURN width * height;
END;
$$ LANGUAGE plpgsql;
```

The CREATE FUNCTION calculate_area(width INTEGER, height INTEGER) statement creates a new function named calculate_area that accepts two integer parameters: width and height. The function returns an integer value, as specified by the RETURNS INTEGER clause. The function body is enclosed within AS $$ ... $$, which is a *dollar-quoted string literal,* allowing for easier readability without needing to escape special characters.

Note   A dollar-quoted string literal in PostgreSQL is a way to define string constants without needing to escape special characters like quotes. It is enclosed within double dollar signs ($$) or custom tags (tag). This is especially useful in functions, stored procedures, and dynamic SQL to improve readability and avoid escaping issues.

Dollar quoting ensures that embedded single quotes don't interfere with SQL syntax, making it easier to write complex queries and function bodies.

The function logic is placed between BEGIN and END, marking the start and end of the executable code block. Inside this block, RETURN width * height; performs the multiplication of the provided width and height values, returning the computed area. The LANGUAGE plpgsql; statement specifies that the function is written in PL/pgSQL, PostgreSQL's procedural language. Once created, the function can be used as follows:

```
SELECT calculate_area(5, 10);
```

This query will return 50, as the area is calculated as 5 * 10.

The following two queries are provided to illustrate the basic syntax of function modification or deletion.

You can update a function by deleting the old version and creating a new one:

```
DROP FUNCTION IF EXISTS function_name;
```

To drop a function, use this statement:

```
DROP FUNCTION function_name;
```

Creating your own functions in PostgreSQL using PL/pgSQL allows you to define complex logic, improve code reusability, and customize your database behavior to fit specific needs. You can handle input/output parameters, error handling, and return values to suit the business logic you are implementing.

Error Handling

Error handling in PostgreSQL is essential for query, function, and transaction reliability. When an error occurs, PostgreSQL aborts the transaction unless it is handled explicitly.

Note   A *transaction* in PostgreSQL is a sequence of SQL statements executed as a single unit of work. It ensures atomicity, meaning all operations either complete successfully or are entirely rolled back if an error occurs. Transactions start with BEGIN, and changes are saved using COMMIT. If an error occurs, ROLLBACK undoes all the changes, maintaining database integrity. In general, transactions in PostgreSQL are typically used for data integrity and consistency in write operations like INSERT, UPDATE, and DELETE, but they can also be used in data analysis tasks when necessary. While pure data analysis usually involves SELECT queries, aggregations, and reporting without requiring transactions, certain situations may benefit from them. PostgreSQL is distinct from many other database engines due to its write-ahead log (WAL) mechanism, which enables rollback of even DDL (Data Definition Language) statements such as CREATE, ALTER, and DROP. This behavior further strengthens its transactional consistency and reliability.

There are several causes of errors, including invalid input, constraints violations, and system failures. PostgreSQL provides exception handling mechanisms to manage such situations, particularly within functions and stored procedures. The RAISE EXCEPTION statement allows developers to generate custom error messages, while the EXCEPTION block enables structured error handling. Proper error management helps prevent unexpected failures, rolls back transactions safely, and provides meaningful feedback to users and applications.

Basic Exception Handling

A function can use the EXCEPTION block to catch and respond to errors. Here is an example that prevents division by zero errors:

```
CREATE FUNCTION safe_divide(numerator INTEGER, denominator INTEGER)
RETURNS FLOAT AS $$
BEGIN
    IF denominator = 0 THEN
        RAISE EXCEPTION 'Error: Division by zero is not allowed';
    ELSE
        RETURN numerator::FLOAT / denominator;
    END IF;
END;
$$ LANGUAGE plpgsql;
```

This function named safe_divide is designed to perform division operations while preventing the common error of dividing by zero. The function accepts two integer parameters: numerator and denominator. When called, it first checks if the denominator equals 0 using an IF statement.

Note Generally, IF ELSE is a conditional statement that allows a program to execute different blocks of code based on whether a condition is true or false. In PostgreSQL, IF ELSE is used inside functions, procedures, and DO blocks. For instance, the basic syntax is as follows:

```
DO $$
BEGIN
    IF 10 > 5 THEN
        RAISE NOTICE 'Condition is True';
    ELSE
        RAISE NOTICE 'Condition is False';
    END IF;
END $$ LANGUAGE plpgsql;
```

In PostgreSQL, a DO block is an anonymous code block that allows executing procedural logic without creating a function. It is useful for running temporary scripts, testing logic, or performing one-time operations.

If the denominator is zero, the function raises an exception with a descriptive error message stating "Error: Division by zero is not allowed", which prevents the operation from proceeding and alerts the user about the issue. If the denominator is not zero, the function proceeds to perform the division operation by first converting the numerator to a floating-point number (using the ::FLOAT cast notation) and then dividing it by the denominator, returning the result as a floating-point number.

Note In PostgreSQL functions, the :: type cast operator is used to explicitly specify the return type of a value before it is returned. This ensures that the function always returns the correct data type and prevents implicit type conversion issues. Recall that in PostgreSQL, :: is the type cast operator, used to explicitly convert one data type into another. It is an alternative to the CAST() function. But you should use :: in function returns for the following reasons:

1. To ensure that the function returns the correct type.

2. To prevent implicit casting errors when dealing with mixed data types.

3. To clarify the type of expected output by explicitly stating it.

Using this method ensures more precise results than dividing by integers. The function is written in PostgreSQL's procedural language (PL/pgSQL), which extends standard SQL by supporting control structures and error-handling features not available in regular SQL.

Using the EXCEPTION Block for Error Handling

The EXCEPTION WHEN clause catches specific errors and defines alternative actions.

```
CREATE FUNCTION insert_employee(name TEXT, age INTEGER)
RETURNS VOID AS $$
BEGIN
    INSERT INTO employees (employee_name, employee_age) VALUES (name, age);
EXCEPTION
    WHEN unique_violation THEN
        RAISE NOTICE 'Employee with the same name already exists.';
    WHEN check_violation THEN
        RAISE NOTICE 'Age must be a positive number.';
    WHEN others THEN
        RAISE NOTICE 'An unknown error occurred.';
END;
$$ LANGUAGE plpgsql;
```

The insert_employee function creates a record for an employee and handles common errors. The function takes two parameters: name, a text string representing the employee's name, and age, an integer value for the employee's age. When executed, it attempts to insert these values into the Employees table, specifically into the employee_name and employee_age columns. This function is particularly robust due to its comprehensive handling of exceptions. If the insertion violates a unique constraint, likely indicating that an employee with the same name already exists in the database, it captures the unique_violation error and displays a notice to the user. Similarly, if the age parameter fails a check constraint (probably requiring that age be positive), it catches the check_violation error and informs the user that age must be positive. Finally, it includes a fallback exception handler for any other unexpected errors that might occur during the insertion process. This function returns VOID, meaning it doesn't return any value but simply performs the insertion operation with appropriate error handling. Table 10-12 summarizes the different error types and use cases for handling exceptions.

Table 10-12. *Error types and Use Cases for Handling Exceptions*

Error Type	Description	Use Case
unique_violation	Catches errors related to inserting duplicate values in columns with a UNIQUE constraint.	Prevents inserting duplicate primary keys or unique values.
check_violation	Handles errors when CHECK constraints, for instance negative values are violated.	Ensures that values follow predefined rules for instance, age must be positive.
others	Catches any unhandled errors that don't match specific types.	Prevents crashes by handling unexpected errors gracefully.

The First Story: Online Clothing Market

Mary is a data analyst at Fashion Flow, an online clothing marketplace. Her job is to analyze sales, customer behavior, and inventory using SQL. As shown in Tables 10-13, 10-14, and 10-15, the company's database contains tables for orders, customers, and products. Mary needs to answer several business questions and perform a few tasks:

- What is the total revenue generated and the average order value?

- How many sales occurred each month?

- Which product has been ordered the most?

- Who are the top customers by spending?

- What are the email domains of customers?

- What are the colors of the available products?

- Ensure that if stock data is missing, it shows "Out of Stock."

- Can Mary create a function that assigns a "loyalty score" based on the total amount a customer has spent?

- What PostgreSQL version is running, and who is the current user?

Table 10-13. *The Customers Table*

customer_id	Name	Email	signup_date	total_spent
1	Alice	alice@mail.com	2022-03-15	300.5
2	Bob	bob@mail.com	2021-07-10	150
3	Carol	carol123@mail.com	2023-01-20	450.75
4	David	dave99@mail.com	2021-12-05	1200
5	Eve	eve_wonder@mail.com	2022-08-25	550.25

Table 10-14. *The Products Table*

product_id	Name	Category	Price	Stock	Attributes (JSON)
101	T-Shirt	Clothing	20	100	{"color": "red", "size": "M"}
102	Jeans	Clothing	50	50	{"color": "blue", "size": "L"}
103	Jacket	Clothing	100	30	{"color": "black", "size": "XL"}
104	Sneakers	Footwear	80	40	{"color": "white", "size": "10"}
105	Hoodie	Clothing	60	25	{"color": "gray", "size": "L"}

Table 10-15. *The Orders Table*

order_id	customer_id	product_id	Quantity	order_date	total_price
1	1	101	2	2023-02-05	40
2	2	102	1	2023-02-10	50
3	3	103	1	2023-02-15	100
4	4	104	2	2023-02-20	160
5	5	105	1	2023-02-25	60
6	1	102	1	2023-03-01	50
7	3	101	3	2023-03-05	60
8	4	105	2	2023-03-10	120

The following query uses aggregate functions to calculate the total revenue generated and the average order value:

```
SELECT
    SUM(total_price) AS total_revenue,
    AVG(total_price) AS avg_order_value
FROM orders;
```

Table 10-16 illustrates the total revenue generated and the average order value.

Table 10-16. *The Total Revenue Generated and the Average Order Value*

total_revenue	avg_order_value
640	80.00000

The following query calculates sales trends per month:

```
SELECT
    DATE_TRUNC('month', order_date) AS month,
    COUNT(order_id) AS total_orders
FROM orders
GROUP BY month
ORDER BY month;
```

Table 10-17 illustrates sales trends per month.

Table 10-17. *Sales Trends Per Month*

Month	total_orders
2023-02-01	5
2023-03-01	3

The following query finds the best-selling products.

```
SELECT
    product_id,
    SUM(quantity) AS total_sold,
    RANK() OVER (ORDER BY SUM(quantity) DESC) AS rank
FROM orders
GROUP BY product_id;
```

Table 10-18 illustrates the best-selling products.

Table 10-18. *Best-Selling Products*

product_id	total_sold	Rank
101	5	1
105	3	2
104	2	3
102	2	3
103	1	5

The following query uses SUM() and RANK() functions to identify top spending customers.

```
SELECT
    customer_id,
    SUM(total_price) AS total_spent,
    RANK() OVER (ORDER BY SUM(total_price) DESC) AS rank
FROM orders
GROUP BY customer_id;
```

Table 10-19 illustrates the top spending customers.

Table 10-19. *Top Spending Customers*

customer_id	total_spent	Rank
4	280	1
3	160	2
1	90	3
5	60	4
2	50	5

To extract the email domains of customers, the following query uses the SPLIT_PART() function:

```
SELECT
    name,
    email,
    SPLIT_PART(email, '@', 2) AS domain
FROM customers;
```

Table 10-20 illustrates the extracted email domains of customers.

Table 10-20. *The Extracted Email Domains of Customers*

Name	Email	Domain
Alice	alice@mail.com	mail.com
Bob	bob@mail.com	mail.com
Carol	carol123@mail.com	mail.com
David	dave99@mail.com	mail.com
Eve	eve_wonder@mail.com	mail.com

To extract product colors, the following query uses the JSON function.

```
SELECT
    name,
    attributes->>'color' AS color
FROM products;
```

Table 10-21 illustrates the extracted product colors.

Table 10-21. *The Extracted Product Colors*

Name	Color
T-Shirt	red
Jeans	blue
Jacket	black
Sneakers	white
Hoodie	gray

To handle missing stock data with default values, the following query uses the control functions to ensure that "Out of Stock" is displayed when stock data is missing:

```
SELECT
    name,
    COALESCE(CAST(stock AS TEXT), 'Out of Stock') AS stock_status
FROM products;
```

Table 10-22 illustrates the stock data values.

Table 10-22. *Stock Data Values*

Name	stock_status
T-Shirt	100
Jeans	50
Jacket	30
Sneakers	40
Hoodie	25

To calculate customer loyalty scores with specific defined functions, the following query creates a function that assigns a "loyalty score" based on the total amount a customer has spent.

First, Mary defines a `calculate_loyalty_score` function that categorizes customers based on their total spending.

```
CREATE FUNCTION calculate_loyalty_score(total_spent NUMERIC)
RETURNS TEXT AS $$
BEGIN
    IF total_spent >= 1000 THEN
        RETURN 'Platinum';
    ELSIF total_spent >= 500 THEN
        RETURN 'Gold';
    ELSIF total_spent >= 200 THEN
        RETURN 'Silver';
    ELSE
        RETURN 'Bronze';
    END IF;
END;
$$ LANGUAGE plpgsql;
```

Then, she uses it in a query to classify the customers:

```
SELECT
    name,
    total_spent,
    calculate_loyalty_score(total_spent) AS loyalty_level
FROM customers;
```

This function automates customer classification, helping the business offer personalized promotions based on spending levels, as illustrated in Table 10-23.

Table 10-23. *Customer Classification*

Name	total_spent	loyalty_level
Alice	300.5	Silver
Bob	150	Bronze
Carol	450.75	Silver
David	1200	Platinum
Eve	550.25	Gold

Lastly, to get the database version and the current user, the following query uses system functions. The results are shown in Table 10-24.

```
SELECT version(), current_user;
```

Table 10-24. *The Database Version and Current User*

Version	current_user
PostgreSQL 14.17 (Debian 14.17-1.pgdg120+1) on x86_64-pc-linux-gnu, compiled by gcc (Debian 12.2.0-14) 12.2.0, 64-bit	user_43c2bv266_43c9rcy88

Breaking Down Complex Problems with Analytical Tools

To turn data into gold with SQL, mastering an advanced set of analytical tools called *recursive queries* is essential. Recursive queries allow you to break down complex problems into smaller, more manageable steps. *Recursive common table expressions* (RCTEs) are especially useful for hierarchical data, such as analyzing organizational structures or network flows. These functions enable analysts to construct a coherent story line around their data, uncovering insights that help them make informed decisions.

The combination of CTEs and recursive queries enables you to structure queries modularly and readably. The use of CTEs allows you to break down complex problems into smaller, more manageable steps, improving query clarity and efficiency. Rather than writing nested subqueries, you can define intermediate results using CTEs and reference them in the main query, making SQL queries more understandable and maintainable. This modular approach allows each step to be reviewed independently, improving debugging and optimization.

It has been mentioned in earlier chapters that a CTE is a temporary result set that can be referenced within a query. As a result, queries are more readable and structured, and large queries can be broken into smaller steps.

```
WITH cte_name AS (
    SELECT column1, column2
    FROM table_name
    WHERE condition
)
SELECT * FROM cte_name;
```

Here, a CTE named cte_name creates a temporary result set. The CTE selects column1 and column2 from table_name where a specific condition is met. The main query then selects all columns from this temporary result set. Accordingly, as an example, the following query finds the total sales per product category from a sales table. By using CTE, it makes it easier to read and maintain.

```
WITH category_sales AS (
    SELECT category, SUM(sales) AS total_sales
    FROM sales_data
    GROUP BY category
)
SELECT * FROM category_sales;
```

Here, the CTE category_sales calculates the total sales per category. To make the SQL easier to read and maintain, the main query simply retrieves the results from category_sales.

A recursive CTE is a special type of CTE that refers to itself. This allows it to process hierarchical or sequential data, such as organizational structures, paths in networks, or parent-child relationships. It consists of two key parts: a base query and a recursive query. The *base query* is the initial dataset, and the *recursive query* calls the CTE itself, repeatedly fetching the next level of data.

```
WITH RECURSIVE cte_name AS (
    -- Base query: Selects the starting point
    SELECT column1, column2
    FROM table_name
    WHERE condition

    UNION ALL

    -- Recursive query: Calls itself to fetch related data
    SELECT t.column1, t.column2
    FROM table_name t
    JOIN cte_name c ON t.related_column = c.column1
)
```

This query employs a RECURSIVE CTE with two main parts. The base query establishes the starting point by selecting column1 and column2 from table_name where a specific condition is met. As a result, this forms the first level of results. The recursive part recursively joins the table back to the previous results, which are stored in the CTE itself. It uses the relationship between t.related_column and c.column1, gradually building up multiple levels of connected data. Each iteration adds new rows until no matching rows are found. This technique is particularly useful for traversing tree structures like organizational hierarchies, parts explosions, or network paths without knowing their depth in advance. The recursion continues until it reaches a natural termination point, where no new rows are added. As an example, the following query wants to find all employees under a specific manager from an employee's table with id, name, and manager_id.

```
WITH RECURSIVE employee_hierarchy AS (
    -- Base case: Select top-level manager (CEO)
    SELECT id, name, manager_id, 1 AS level
    FROM employees
    WHERE manager_id IS NULL

    UNION ALL

    -- Recursive case: Find employees reporting to the previous level
    SELECT e.id, e.name, e.manager_id, eh.level + 1
    FROM employees e
    JOIN employee_hierarchy eh ON e.manager_id = eh.id
)
```

This query starts with the CEO and recursively finds all employees under them, assigning hierarchical levels dynamically. This query uses a recursive CTE called employee_hierarchy to generate a hierarchical view of an organization's reporting structure. The query consists of two main components that work together to build the complete organizational hierarchy. The base case establishes the starting point by selecting the top-level manager, typically the CEO, from the Employees table, identified by having a NULL value in the manager_id column. This query selects the manager's id, name, and manager_id, while also assigning a level value of 1 to indicate the top position in the hierarchy. This is the first row of the result set and serves as the basis for recursion.

The recursive part then joins the Employees table with the previously generated results, stored in the employee_hierarchy CTE. It finds all employees whose manager_id matches the ID of someone in the current result set, essentially identifying all direct reports. For each of these employees, it increases the level value by 1, indicating they are one level lower in the hierarchy than their manager. This recursive process continues automatically, with each iteration finding employees who report to managers discovered in the previous iteration. This continues until no more matching employees are found.

This query efficiently traverses the entire organizational structure regardless of its depth using this recursive approach. This results in a complete picture of the reporting relationships with appropriate level indicators. This makes it possible to visualize complex organizational hierarchies without having to know in advance how many levels there are.

CTEs and recursive CTEs are powerful tools that make SQL queries more structured, readable, and efficient. Table 10-25 illustrates the key differences between CTEs and RCTEs.

Table 10-25. *Key Differences between CTEs and RCTEs*

Feature	Standard CTE	Recursive CTE
Purpose	Organizes complex queries	Works with hierarchical or sequential data
Calls itself	No	Yes
Use case	Filtering, grouping, aggregations	Organizational hierarchies, graph traversal
Query execution	Runs once	Runs iteratively until no more rows match

The Second Story: An Analysis of a Family Tree for the Civil Registration Office

The civil registration process is where records of life events—such as births, marriages, and deaths—are kept for the citizens and residents of a city. Jane intends to work in the ancestry research department, where she will be analyzing data collected from these records. Jane's task is to analyze a family tree dataset. The dataset includes information about individuals and their direct parents. Using RCTEs, she will solve various challenges related to family lineage. For Jane's analytic journey, Table 10-26 contains data associated with citizens and residents.

Table 10-26. *The family_tree Table*

ID	Name	parent_id	birth_year
1	Alice	NULL	1950
2	Bob	1	1975
3	Carol	1	1978
4	Dave	2	2000
5	Eve	3	2003
6	Frank	2	2005
7	Grace	5	2028

Given this data, Jane wants to accomplish the following:

- Retrieve all descendants of Alice along with their generational depth.

- List all ancestors of Grace, tracing back through generations.

- Identify the longest ancestral chain in the family tree.

- Find the age difference between the oldest ancestor and youngest descendant for each lineage.

The following query will retrieve all descendants of Alice together with their generational depth:

```
WITH RECURSIVE descendants AS (
    -- Base case: Start with Alice
    SELECT id, name, parent_id, birth_year, 1 AS generation
    FROM family_tree
    WHERE name = 'Alice'

    UNION ALL

    -- Recursive case: Find children of previous generation
    SELECT f.id, f.name, f.parent_id, f.birth_year, d.generation + 1
    FROM family_tree f
    JOIN descendants d ON f.parent_id = d.id
)
SELECT * FROM descendants;
```

To find all descendants of Alice within a family tree, this SQL query constructs a recursive query using the WITH RECURSIVE clause. This query starts by selecting all data from the family_tree table, including information about the current ancestor of Alice, and then iteratively calls itself, the descendants CTE, to find all children of that ancestor. This CTE will join each row from the family_tree table to the Descendants table based on the parent_id, progressively exploring the family tree and ultimately returning a list of all descendants of Alice, including their IDs, names, parent IDs, and birth years.

Note Part of the WITH RECURSIVE clause defines a new column called generation within the descendants CTE. It plays a crucial role in tracking the "level" or "depth" of each individual's ancestry within the family tree. In the previous query, the 1 AS generation part declares and names a column called generation that will be added to every row produced by the recursive CTE. The value 1 initially assigned to this column for the base case (Alice) establishes the starting point for counting generational depth. As the query recurs, moving down through the family tree, the generation value increases by one for each generation.

Alice is in generation one, so she's the starting point. Bob's children will be in generation two, because they're one generation removed from Alice. Generation 3 refers to Alice's grandchildren. The generation column allows you to organize and visualize the family tree. Using parental lineage, it builds a numerical hierarchy of ancestry relationships.

Table 10-27 illustrates all descendants of Alice along with their generational depth.

Table 10-27. *Descendants of Alice and Their Generational Depth*

ID	Name	parent_id	birth_year	generation
1	Alice	NULL	1950	1
2	Bob	1	1975	2
3	Carol	1	1978	2
4	Dave	2	2000	3
5	Eve	3	2003	3
6	Frank	2	2005	3
7	Grace	5	2028	4

The following query provides a list of all the ancestors of Grace, tracing back through the generations:

```
WITH RECURSIVE ancestors AS (
    -- Base case: Start with Grace
    SELECT id, name, parent_id, 1 AS generation
    FROM family_tree
    WHERE name = 'Grace'

    UNION ALL

    -- Recursive case: Find parents of previous generation
    SELECT f.id, f.name, f.parent_id, a.generation + 1
    FROM family_tree f
    JOIN ancestors a ON f.id = a.parent_id
)
SELECT * FROM ancestors;
```

This query generates a recursive table named Ancestors. It begins with a base case selecting all records from the family_tree table representing the starting point, Grace, with her ID, name, parent ID, and a generation of 1. Then, this query iteratively calls itself, the ancestors CTE, to find all the parents of the previous generations, starting from Grace. It uses the parent ID from each row in the current generation to find the parents of the current generation. Finally, the query selects all rows returned by the recursive call, effectively creating a table of all the parents within the family tree. This shows a complete view of the ancestors, as shown in Table 10-28.

Table 10-28. *All Ancestors of Grace*

ID	Name	parent_id	Generation
7	Grace	5	1
5	Eve	3	2
3	Carol	1	3
1	Alice	NULL	4

The following query identifies the longest ancestral chain in the family tree to determine the longest lineage:

```
WITH RECURSIVE lineage AS (
    -- Base case: Start with individuals who have no parents
    SELECT id, name, parent_id, 1 AS depth
    FROM family_tree
    WHERE parent_id IS NULL

    UNION ALL

    -- Recursive case: Count generations for each descendant
    SELECT f.id, f.name, f.parent_id, l.depth + 1
    FROM family_tree f
    JOIN lineage l ON f.parent_id = l.id
)
SELECT * FROM lineage ORDER BY depth DESC LIMIT 1;
```

This SQL query uses an RCTE named `lineage` to traverse a family tree structure stored in a table called `family_tree`. It starts with the base case of finding individuals who have no parents, where `parent_id` is NULL, and assigns them a depth of 1. The recursive part then joins these results to the `family_tree` table to find their children, adding one to the depth counter for each generation. This process continues building the lineage hierarchy until all descendants are found. The final SELECT statement sorts the results by depth in descending order and limits the output to just one row. Essentially, this identifies the youngest descendant in terms of generational depth in the family tree with the greatest distance from the original ancestors. Table 10-29 identifies the longest ancestral chain in the family tree.

Table 10-29. *The Longest Ancestral Chain in the Family Tree*

ID	Name	parent_id	Depth
7	Grace	5	4

The following query aims to find the age gap between the oldest and youngest in each lineage:

```
WITH RECURSIVE family_age AS (
    -- Base case: Start with individuals who have no parents
    SELECT id, name, parent_id, birth_year, birth_year AS oldest_birth_year
    FROM family_tree
    WHERE parent_id IS NULL

    UNION ALL

    -- Recursive case: Compare birth years through generations
    SELECT f.id, f.name, f.parent_id, f.birth_year, fa.oldest_birth_year
    FROM family_tree f
    JOIN family_age fa ON f.parent_id = fa.id
)
SELECT
    name,
    birth_year,
```

```
        oldest_birth_year,
        birth_year - oldest_birth_year AS age_gap
FROM family_age
ORDER BY age_gap DESC;
```

This SQL query uses a RCTE named `family_age` to analyze age relationships within a family tree. Starting with individuals who have no parents, *root ancestors,* as the base case, it captures their ID, name, parent ID, birth year. It establishes their birth year as the `oldest_birth_year` reference point. The recursive part then joins these results to the `family_tree` table to add descendants. It maintains the original ancestor's birth year as it traverses through generations. After building this complete family structure with preserved ancestral birth years, the final `SELECT` statement calculates the `age_gap` by subtracting the oldest ancestor's birth year from each family member's birth year, effectively measuring the time span between generations. The results are sorted by age gap in descending order. As a result, family members with the largest generational age differences are highlighted first, providing insight into how the age spans of the generations have changed over time. Table 10-30 shows the age difference between the oldest ancestor and youngest descendant for each lineage.

Table 10-30. *The Age Difference Between the Oldest Ancestor and Youngest Descendant for Each Lineage*

Name	birth_year	oldest_birth_year	age_gap
Grace	2028	1950	78
Eve	2003	1950	53
Frank	2005	1950	55

Summary

This chapter highlighted SQL's functions in enhancing data analysis and insight extraction. It explored techniques for manipulating and analyzing data using aggregate functions for summarization, statistical and mathematical functions for numerical insights, and window functions for trend analysis. Additionally, value, ranking, string, date and time, JSON, control, and system functions allow for deeper data exploration and transformation. Moreover, SQL simplifies insightful summaries and hierarchical data analysis. These techniques collectively improve overall data understanding, making complex datasets more accessible and actionable.

Key Points

- SQL functions enhance data analysis by enabling powerful calculations, filtering, and transformations for deeper insights.

- The combination of recursive queries and CTEs breaks down complex problems into manageable steps, enabling hierarchical data analysis.

- Custom SQL functions allow users to extend SQL's capabilities by defining reusable logic for specialized calculations.

- With these SQL functions, data workflows become more efficient, enabling better decision-making and actionable insights.

Key Takeaways

- **A variety of data processing**: SQL functions provide powerful tools for aggregating, filtering, and transforming data to uncover deeper insights.

- **Advanced analytical capabilities**: Utilize statistical, mathematical, and window functions to perform complex calculations.

- **Efficient hierarchical analysis**: Recursive common table expressions (RCTEs) help break down complex relationships.

- **Data structure processing**: Take advantage of JSON, string, and date functions to manage and manipulate structured and semi-structured data effectively.

- **Custom functionality expansion**: Define reusable SQL functions to encapsulate logic and streamline repetitive calculations for more efficient workflows.

Looking Ahead

The next chapter, "The Grand Finale: Presenting Your Data Story," examines how to present your data story in the best way possible. Some closing thoughts are also shared. This includes effective techniques for data story presentation, and tips that can enhance data interpretation by creating reusable, efficient analytical workflows.

Test Your Skills

Emma is a data analyst at Scholars Academy, where students from different grades participate in an annual competition called the Excellence Competition. The school administration needs insights into student performance across different subjects and categories. They ask Emma to answer these questions based on the data in Table 10-31:

1. Which students performed best in each subject, ranking them within their grade levels?

2. What is the average score per subject, and how does it compare across different grades?

3. What percentage of students scored above 90 percent (Excellent) and below 50 percent (Needs Improvement)?

4. How do students' scores in core subjects (Mathematics, Science) vary across different competitions?

5. What proportion of students improved their scores in the same subject compared to last year's competition?

6. Which students achieved the highest total scores across all subjects?

7. How many students participated in each grade, and what was the median score per grade?

8. What is the highest, lowest, and average score per subject?

9. Is it possible to create a function that categorizes students in accordance with their performance level (Excellent, Good, Average, Needs Improvement)?

Table 10-31. *The competition_results Table*

student_id	Name	Grade	Subject	Score	competition_year
101	Alice	9	Mathematics	92	2024
102	Bob	10	Science	78	2024
103	Charlie	9	Mathematics	85	2024
104	David	11	Literature	65	2024
105	Emma	10	Science	88	2024

CHAPTER 11

The Grand Finale: Presenting Your Data Story

Your journey is nearing its end. The journey continued throughout each chapter with stories involving data analysis using SQL queries. Through chapter-by-chapter stories, you have learned how to extract deeper insights and information from raw data to address complex and advanced analytical questions. This chapter aims to summarize the previous chapters and provide insight into presenting a narrative for data analysis. Up until this chapter, the chapters contributed to teaching data analysis and SQL query writing through narratives. However, this chapter focuses more on providing a mindset for readers who have acquired sufficient query writing skills in data analysis.

The Art of Data Storytelling

Storytelling in data analysis is the practice of presenting data-driven insights in a compelling and understandable way. Instead of just showing raw numbers, tables, or charts, storytelling helps communicate what the data means and why it matters. It turns complex findings into a narrative that people can relate to, making it easier to understand trends, patterns, and conclusions. Data storytelling consists of four essential components: data, narrative, visuals, and actionable insights. Data consists of facts, numbers, and statistical results, which provide the foundation for storytelling. Narrative is a structured explanation that gives context to the data, making it meaningful. Visuals like graphs, charts, and dashboards support and enhance the story. Lastly, actionable insights are conclusions that guide decision-making or strategic planning.

© Hamed Tabrizchi 2025
H. Tabrizchi, *Narrative SQL*, https://doi.org/10.1007/979-8-8688-1560-7_11

The Importance of Query Writing for Storytelling in Data Analysis

In data storytelling, SQL plays a crucial role in extracting, transforming, and analyzing data to create meaningful narratives. Well-written SQL queries allow analysts to retrieve relevant data efficiently, uncover insights, and present findings in a structured manner. Without strong query-writing skills, data storytelling lacks accuracy, depth, and clarity.

Extracting the Right Data for a Story

The foundation of storytelling in data analysis starts with retrieving relevant and accurate data. SQL enables analysts to filter and organize data, ensuring that only the necessary information is included in the analysis. Table 11-1 provides a structured table summarizing SQL statements, clauses, operations, and functions used to extract the right data for storytelling.

Table 11-1. *SQL Statements, Clauses, Operations, and Functions Used to Extract the Right Data for Storytelling*

Category	SQL Components	Purpose
SQL statements	SELECT	Retrieves specific data from tables
	FROM	Specifies the table(s) to query
	WHERE	Filters data based on conditions
	GROUP BY	Aggregates data by specific columns
	ORDER BY	Sorts results in ascending or descending order
	JOIN	Combines data from multiple tables
	LIMIT	Restricts the number of returned rows
	HAVING	Filters aggregated results

(continued)

Table 11-1. *(continued)*

Category	SQL Components	Purpose
SQL clauses	DISTINCT	Removes duplicate values
	AS	Renames columns for better readability
	IN / NOT IN	Matches values in a specified list
	BETWEEN	Filters data within a range
	LIKE	Searches for patterns in text
SQL operations	Arithmetic (+, -, *, /)	Performs mathematical calculations
	Comparison (=, >, <, >=, <=, <>)	Compares values
	Logical (AND, OR, NOT)	Combines multiple conditions for filtering
Aggregate functions	COUNT()	Counts the number of rows
	SUM()	Calculates the total sum of values
	AVG()	Computes the average of a column
	MIN() / MAX()	Finds smallest or largest values
String functions	UPPER() / LOWER()	Converts text to uppercase or lowercase
	SUBSTRING()	Extracts part of a string
	TRIM()	Removes extra spaces from text
Date functions	NOW()	Returns the current timestamp
	DATE_TRUNC()	Rounds date/time values to a specified level
	EXTRACT()	Retrieves year, month, or day from a date

Data storytelling relies on SQL to extract and shape precise data from databases. As explained in Chapters 2-6, SQL statements like SELECT and FROM identify the desired data columns and their source tables, respectively. Filtering is achieved using WHERE for row-level conditions and HAVING for conditions on aggregated results produced by GROUP BY. Data arrangement is handled by ORDER BY for sorting, while JOIN integrates data from multiple tables. LIMIT controls output volume.

Further refinement comes from SQL clauses. DISTINCT ensures uniqueness in results, AS renames columns for clarity, and clauses like IN, NOT IN, BETWEEN, and LIKE offer powerful pattern-matching and range-filtering capabilities. SQL operations, including arithmetic (+, -, *, /), comparison (=, >, <), and logical (AND, OR, NOT), allow for calculations and complex conditional logic within queries. As discussed in detail in Chapters 5 and 10, aggregate functions such as COUNT(), SUM(), AVG(), MIN(), and MAX() are critical for summarizing data insights. Additionally, string functions like UPPER(), SUBSTRING(), and TRIM() and date functions like NOW(), DATE_TRUNC(), and EXTRACT() facilitate text manipulation and temporal analysis, ensuring the retrieved data effectively supports the narrative.

Structuring Data for Better Insights

Raw datasets are often complex and difficult to interpret. SQL helps structure data by using JOINs, aggregations, and filtering, making insights easier. See Table 11-2.

Table 11-2. *SQL Statements, Clauses, Operations, and Functions Used to Structure Data for Better Insights*

Category	SQL Components	Purpose
SQL statements	SELECT	Extracts specific columns from a table
	FROM	Defines the table(s) to retrieve data from
	WHERE	Filters data based on conditions
	GROUP BY	Groups data to perform aggregations
	ORDER BY	Sorts the data in ascending or descending order
	JOIN	Merges data from multiple tables
	LIMIT	Restricts the number of rows returned
	HAVING	Filters aggregated data

(continued)

Table 11-2. (*continued*)

Category	SQL Components	Purpose
SQL clauses	DISTINCT	Removes duplicate values
	AS	Renames columns for better readability
	IN / NOT IN	Filters values based on a predefined list
	BETWEEN	Filters values within a specified range
	LIKE	Finds text patterns using wildcards
SQL operations	Arithmetic (+, -, *, /)	Performs mathematical calculations
	Comparison (=, >, <, >=, <=, <>)	Compares values
	Logical (AND, OR, NOT)	Combines multiple filter conditions
Aggregate functions	COUNT()	Counts the number of rows
	SUM()	Calculates the total sum of values
	AVG()	Computes the average of a column
	MIN() / MAX()	Finds the smallest or largest values
String functions	CONCAT()	Combines multiple string values
	UPPER() / LOWER()	Changes text to uppercase or lowercase
	SUBSTRING()	Extracts a portion of a string
	TRIM()	Removes extra spaces from text
Date functions	NOW()	Returns the current timestamp
	DATE_TRUNC()	Rounds date/time values
	EXTRACT()	Retrieves specific parts of a date
Window functions	ROW_NUMBER()	Assigns a unique row number
	RANK() / DENSE_RANK()	Assigns ranking to rows based on criteria
	LAG() / LEAD()	Accesses previous or next row values
	NTILE(n)	Divides data into n equal parts
Common table expressions (CTEs)	CASE WHEN	Performs conditional calculations
	WITH (Common Table Expressions)	Creates temporary result sets for better readability

As shown in Table 11-2, SQL provides essential components to structure data for better insights. Statements like SELECT, FROM, WHERE, JOIN, and GROUP BY retrieve, filter, combine, and aggregate information. Clauses such as AS, LIKE, and BETWEEN refine queries. Operations enable calculations and logic, while the aggregate, string, and date functions summarize and format data. As discussed in Chapter 10, advanced features such as RANK and LAG perform calculations across related rows. Also, CASE WHEN and CTEs, discussed in Chapter 8, provide conditional logic and improve query organization, resulting in more insightful and structured results.

Supporting Visualizations of Query Results

SQL queries provide clean and structured data for dashboards, charts, and reports. Well-written queries ensure accurate, relevant, and insightful visualizations. Table 11-3 lists the SQL elements that can help you prepare query results for dashboards, reports, and visual storytelling.

Table 11-3. *SQL Statements, Clauses, Operations, and Functions Used to Support Visualizations of Query Results*

Category	SQL Components	Purpose
SQL statements	SELECT	Retrieves specific data for visualization
	FROM	Specifies the table(s) to query
	WHERE	Filters data before visualization
	GROUP BY	Aggregates data for charts (e.g., bar, pie)
	ORDER BY	Sorts data for trend analysis
	JOIN	Merges data from multiple sources
	LIMIT	Restricts data size for better visualization
	HAVING	Filters aggregated data for clarity
SQL clauses	DISTINCT	Removes duplicate values to avoid redundancy
	AS	Renames columns for clearer labels
	IN / NOT IN	Selects specific values for focused analysis
	BETWEEN	Filters data within a defined range
	LIKE	Searches for pattern-matching text

(continued)

Table 11-3. (*continued*)

Category	SQL Components	Purpose
SQL operations	Arithmetic (+, -, *, /)	Computes numerical values for visual analysis
	Comparison (=, >, <, >=, <=, <>)	Filters data for meaningful visualization
	Logical (AND, OR, NOT)	Combines multiple conditions for segmentation
Aggregate functions	COUNT()	Counts data points for histograms or KPIs
	SUM()	Computes totals for bar and pie charts
	AVG()	Calculates averages for trend analysis
	MIN() / MAX()	Identifies range for visual scales
String functions	CONCAT()	Combines values for label formatting
	UPPER() / LOWER()	Standardized text case for consistency
	SUBSTRING()	Extracts key text parts for visualization
	TRIM()	Cleans data for better presentation
Date functions	NOW()	Fetches the current timestamp for timelines
	DATE_TRUNC()	Aggregates time data for trends (e.g., monthly sales)
	EXTRACT()	Retrieves year, month, or day for charts
Window functions	ROW_NUMBER()	Assigns unique row numbers for ranking
	RANK() / DENSE_RANK()	Helps create leaderboards and sorted lists
	LAG() / LEAD()	Compares current vs. previous data points
	NTILE(n)	Splits data into quantiles for distribution analysis
Common Table Expressions (CTEs)	CASE WHEN	Enables conditional categorization in charts
	WITH (CTEs)	Prepares structured data for visualization tools

Chapter 8 explains how SQL prepares results for clear visualization. Statements like SELECT, FROM, WHERE, GROUP BY, and ORDER BY extract, filter, aggregate, and sort data specifically for charts and trend analysis. Clauses such as AS provide clear labels, while operations and aggregate functions compute values essential for visual representation. String and date functions format text and group time-based data. The use of advanced tools such as window functions and CTE provides a variety of visual comparisons, rankings, and categorizations for complex data.

Ensuring Data Accuracy and Reliability

Poorly written queries can lead to incorrect insights, misleading decision-making, and flawed storytelling. As shown in Table 11-4, SQL abilities such as data validation, error handling, and consistency checks ensure trustworthy data in a story.

Table 11-4. SQL Statements, Clauses, Operations, and Functions Used to Ensure Data Accuracy and Reliability

Category	SQL Components	Purpose
SQL statements	SELECT	Retrieves data to verify accuracy
	INSERT	Ensures correct data entry
	UPDATE	Modifies existing data while maintaining integrity
	DELETE	Removes incorrect or duplicate records safely
	MERGE	Prevents duplication by inserting or updating data
	TRANSACTION	Ensures atomicity, consistency, and rollback if needed

(*continued*)

Table 11-4. (*continued*)

Category	SQL Components	Purpose
SQL clauses	WHERE	Filters out incorrect or unwanted data
	HAVING	Ensures aggregated data meets conditions
	ORDER BY	Organizes data to identify inconsistencies
	DISTINCT	Eliminates duplicate records
	CHECK	Defines constraints to maintain data validity
	FOREIGN KEY	Maintains referential integrity
	PRIMARY KEY	Ensures uniqueness of records
	UNIQUE	Prevents duplicate values in specified columns
	NOT NULL	Ensures mandatory fields are not left empty
SQL operations	Arithmetic (+, -, *, /)	Ensures mathematical correctness in calculations
	Comparison (=, >, <, >=, <=, <>)	Validates correct relationships between values
	Logical (AND, OR, NOT)	Combines multiple conditions for accurate filtering
Aggregate functions	COUNT()	Detects missing or duplicate records
	SUM()	Verifies numerical totals
	AVG()	Checks for outliers in average calculations
	MIN() / MAX()	Identifies unexpected values or errors
String functions	TRIM()	Removes unwanted spaces that may cause mismatches
	UPPER() / LOWER()	Standardizes text case to ensure consistency
	LENGTH()	Detects unusually short or long values
	REPLACE()	Fixes incorrect text entries

(*continued*)

Table 11-4. *(continued)*

Category	SQL Components	Purpose
Date functions	NOW()	Captures accurate timestamps
	EXTRACT()	Validates specific date components (year, month, day)
	DATE_TRUNC()	Standardizes date/time for consistency
	AGE()	Checks logical date differences (e.g., age calculation)
Validation functions	IS NULL / IS NOT NULL	Detects missing or incomplete data
	COALESCE()	Replaces NULL values with defaults
	CASE WHEN	Applies conditional corrections to data
	REGEXP_MATCH()	Ensures valid data format using regex
Window functions	ROW_NUMBER()	Identifies duplicate records
	RANK()	Detects inconsistencies in ranked datasets
	LAG() / LEAD()	Compares data to previous or next row
Data integrity tools	CHECKSUM()	Verifies data integrity across tables
	FOREIGN KEY CASCADE	Maintains consistency in relational data

As discussed in the first chapter, a variety of options can be used to ensure the accuracy and reliability of data. Statements like INSERT, UPDATE, and DELETE, governed by the TRANSACTION control, manage data safely. Constraints like PRIMARY KEY, FOREIGN KEY, CHECK, and NOT NULL enforce data validity and integrity rules. Clauses like WHERE filter out inaccuracies, while DISTINCT removes duplicates. Operations and specialized functions like Aggregate, string, date, and validation verify calculations, standardize formats, detect errors, handle NULLs like COALESCE, and validate data.

Automating Storytelling with Dynamic Queries

SQL can automate reports and storytelling by generating dynamic insights that update over time. Stored procedures, views, and scheduled queries enable ongoing analysis, ensuring that decision-makers receive fresh insights regularly, as briefly summarized in Table 11-5.

Table 11-5. *SQL Statements, Clauses, Operations, and Functions Used to Automate Storytelling with Dynamic Queries*

Category	SQL Components	Purpose
SQL statements	SELECT	Retrieves dynamic data for storytelling
	INSERT	Stores generated insights into reporting tables
	UPDATE	Updates dynamic reports based on new data
	DELETE	Removes outdated or irrelevant insights
	WITH (CTE)	Simplifies complex queries for dynamic analysis
	EXECUTE	Runs dynamic queries stored as prepared statements
	PREPARE	Prepares a query with placeholders for dynamic execution
SQL clauses	WHERE	Filters data dynamically based on conditions
	HAVING	Filters aggregated insights dynamically
	ORDER BY	Sorts data dynamically for readability
	GROUP BY	Aggregates data dynamically for summaries
	CASE WHEN	Applies conditional logic for dynamic insights
	LIMIT	Controls result size dynamically
	OFFSET	Supports pagination in automated reports
SQL operations	LIKE / ILIKE	Enables pattern-based dynamic searches
	IN	Filters multiple values dynamically
	BETWEEN	Handles dynamic date and range filters

(*continued*)

Table 11-5. (*continued*)

Category	SQL Components	Purpose
String functions	CONCAT()	Dynamically combines text for reports
	STRING_AGG()	Aggregates multiple values dynamically
	REPLACE()	Adjusts text dynamically
Date functions	NOW()	Uses current timestamps dynamically
	EXTRACT()	Retrieves dynamic date components
	DATE_TRUNC()	Aggregates time-based storytelling insights
	AGE()	Calculates differences for dynamic timelines
Validation functions	COALESCE()	Handles NULL values dynamically
	NULLIF()	Avoids unnecessary errors in dynamic queries
	GREATEST() / LEAST()	Selects dynamic min/max values
Window functions	ROW_NUMBER()	Enables ranking for dynamic comparisons
	RANK()	Generates dynamic rankings
	LAG() / LEAD()	Compares previous and next rows dynamically
Stored procedures	CREATE PROCEDURE	Automates repetitive storytelling queries
	CALL	Executes stored storytelling logic, only supported from PG 11+
Views and materialized views	CREATE VIEW	Stores reusable dynamic queries for reports
	REFRESH MATERIALIZED VIEW	Updates dynamic precomputed insights
Dynamic query execution	FORMAT()	Constructs dynamic SQL queries as strings
	EXECUTE	Runs dynamic SQL queries

SQL facilitates automated storytelling by enabling dynamic queries, as discussed in Chapter 7. Key statements like SELECT and clauses such as WHERE, HAVING, and ORDER BY use nested queries, parameters, and functions like NOW(), EXTRACT(), or CASE WHEN for flexible data retrieval and conditional logic. CTEs and window functions structure complex, context-aware insights. Automation relies on reusable stored procedures and views, as discussed in Chapter 9. Dynamic SQL execution allows queries to generate up-to-date narratives automatically based on changing data or inputs.

Overall, SQL query writing is an essential skill for storytelling in data analysis and it enables analysts to extract meaningful data, structure information for better insights, support visual storytelling with accurate results, ensure reliability and accuracy of data, and automate dynamic reports for continuous storytelling.

PostgreSQL Query Execution

During the execution of a query, PostgreSQL first checks whether it is valid in terms of syntax and structure. This step includes lexical analysis, syntax analysis, and semantic analysis. *Lexical analysis* breaks the query down into individual tokens like keywords, column names, and table names. *Syntax analysis* checks whether the query follows SQL grammar rules. *Semantic analysis* ensures referenced tables or columns exist and verifies data types. Once the query is parsed, PostgreSQL creates an execution plan by considering multiple strategies. PostgreSQL's key actions in these steps include table scanning methods, join strategy selection and sorting and aggregation optimization. In table scanning methods, PostgreSQL decides between a sequential scan that scans the full table, an index scan that uses indexes, and a bitmap heap scan that optimizes scans with multiple indexes. In the join strategy selection, a nested loop join is good for small tables, a hash join is efficient for large datasets, and a merge join is good for sorted data. In sorting and aggregation optimization, PostgreSQL optimizes sorting and aggregation using indexes, hashing, parallel processing, and memory-efficient algorithms to reduce disk I/O and computation time.

Figure 11-1 illustrates the execution order of SQL statements and clauses. In PostgreSQL, the execution order of SQL statements and clauses follows a specific sequence that differs from the order in which they are written in the query. Generally, it begins with the FROM clause, where PostgreSQL identifies the source of the data, which could involve multiple tables or views, and applies any JOIN operations to combine these sources. Next, the WHERE clause filters the rows based on the conditions specified. After filtering, the GROUP BY clause groups the rows into aggregates. If there are any conditions on the aggregated groups, the HAVING clause is then applied. Once the rows are grouped and filtered by HAVING, if applicable, the SELECT clause is used to choose the columns and expressions that will appear in the result. The data is then sorted using the ORDER BY clause, determining the order in which the rows are returned. Finally, LIMIT and OFFSET are applied to restrict the number of rows returned and, in the case of OFFSET, to skip a specified number of rows.

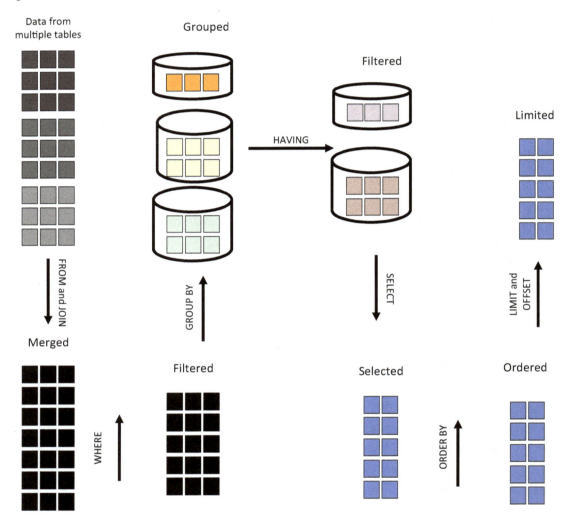

Figure 11-1. *The flow of query execution in PostgreSQL.*

It is essential to understand this execution flow. By understanding this flow of execution in PostgreSQL, you can optimize queries and ensure that they return the right results.

Beyond the Query

The journey of data analysis doesn't end with the execution of a SQL query. While querying serves as the foundation for gathering insights, the ultimate goal is to communicate these insights effectively. Often abstract and complex, raw data requires transformation into a meaningful story. This is where data storytelling comes into play. It acts as the bridge that connects the data's raw form to meaningful understanding. Data analysts can connect trends, data points, and patterns in interesting, engaging, and impactful ways through storytelling.

In order to increase analytical effectiveness, it is essential to present your findings effectively. To get your audience to understand and act on the data, it is not sufficient to simply display the data; the key is to explain the data in a way they can easily understand. A well-crafted presentation can transform raw numbers into a powerful narrative that guides decision-making. By organizing data clearly, using visual aids effectively, and focusing on the insights, it is possible to ensure that your findings are not just seen, but understood. The presentation should not only convey information but also tell a story that highlights the significance of the data. It should also highlight the impact of the conclusions drawn from them.

The previous chapters explored the steps of data analysis, from understanding the problem and preparing the data to analyzing and interpreting it. Now, it's time to shift focus and discuss how to bring the findings to life.

Structuring Your Data Narrative for Presentation

Structuring your data narrative for presentation is crucial to presenting complex information effectively. A well-constructed data presentation most often has a classic storytelling structure, beginning with the presentation.

The classic narrative structure—exposition, rising action, climax, and resolution—can be a strong foundation for data storytelling. This structure helps you make data-driven insights more engaging and memorable by framing them in a meaningful way. In a nutshell, data storytelling involves a structured narrative that inspires action, going beyond mere statistics to convey a compelling message. The classic narrative structure begins with exposition, which sets the stage, where the context, problem, or question at hand is introduced. Providing relevant background information and data is crucial to establish the significance of the topic and help the audience understand why it matters. Following this is rising action, which builds the case, presenting key findings and insights. Data visualizations, trend analyses, and comparisons are employed to build expectation and guide the audience toward the core message. The narrative then reaches its climax, where the most impactful insight or unexpected discovery is revealed. This is the pivotal moment, designed to inspire a significant realization in the audience. Finally, the resolution, which calls to action or next steps, explains the implications of the findings. It suggests possible actions, decisions, or future directions based on the data. This concluding phase ensures that the data story not only informs but also inspires action, transforming raw data into a basis for informed decision-making.

Understanding the Audience

The importance of knowing your audience cannot be underestimated. Customizing your presentation to the audience's proficiency is essential. For data scientists, detailed methodology and technical terminology may be appropriate, while business executives or general stakeholders require a high-level overview that focuses on strategic implications. Further, recognizing your audience's objectives—whether they seek strategic business insights, operational improvements, or technical solutions—will enable you to align your content with their needs. By aligning your narrative with their concerns, you ensure that your narrative addresses the audience's concerns directly and provides them with useful information.

The Key Elements of a Data Story Presentation

The key elements of a data story presentation begin with a powerful intro, designed to immediately capture attention. This could be a thought-provoking question, a startling statistic, or a relatable real-world scenario. Following the opening line, provide context and background, outlining the dataset's origin and the business problem or research question being investigated. A concise description of the methodology, including any SQL queries and analytical techniques used, adds credibility to your findings. The key findings should be presented clearly and concisely, ideally supported by visualizations for enhanced comprehension. Transitioning from findings to practical recommendations is essential. This can translate insights into practical steps stakeholders can implement. This ensures that the presentation concludes with a clear purpose and impact by directing the audience to take action, whether making a business decision, adopting a new strategy, or pursuing further research.

Visualization: Insights from Visuals

Data in its raw form can be dense and difficult to interpret. Visualizations transform raw data into digestible insights, enhancing understanding and engagement. Because the human brain processes visual information more efficiently than textual information, visualizations can be a powerful communication tool. An effective visualization depends, however, on the selection of the right chart type. For example, bar charts are ideal for comparing categories, whereas line charts are excellent for displaying trends over time. A properly chosen chart ensures that the data's story is conveyed accurately and effectively, preventing misinterpretations and enhancing understanding. Table 11-6 and Figure 11-2 illustrate commonly used chart types that are useful for effective visualization.

Table 11-6. *A Summary of Commonly Used Chart Types*

Chart Type	Description
Bar chart	Displays categorical data with rectangular bars; useful for comparing quantities.
Column chart	Similar to a bar chart but with vertical bars; used to compare different categories.
Line chart	Shows trends over time by connecting data points with a line.
Scatterplot	Displays relationships between two numerical variables using dots.
Pie chart	Represents proportions of a whole using slices; best for showing percentage distributions.
Doughnut chart	A variation of the pie chart with a hole in the center, improving readability.
Histogram	Similar to a bar chart but used for frequency distribution of continuous data.
Box plot (box-and-whisker plot)	Summarizes data distribution with quartiles and outliers.

(continued)

Table 11-6. *(continued)*

Chart Type	Description
Bubble chart	Similar to a scatterplot but includes a third variable represented by bubble size.
Heatmap	Uses color gradients to show relationships between variables in a matrix format.
Area chart	Similar to a line chart but with shaded areas under the lines to emphasize volume.
Stacked bar chart	Segments bars into different categories to show part-to-whole relationships.
Waterfall chart	Shows how an initial value is affected by a series of positive or negative changes.
Radar chart (spider chart)	Compares multiple variables across different categories in a radial layout.
Treemap	Uses nested rectangles to represent hierarchical data with size and color.

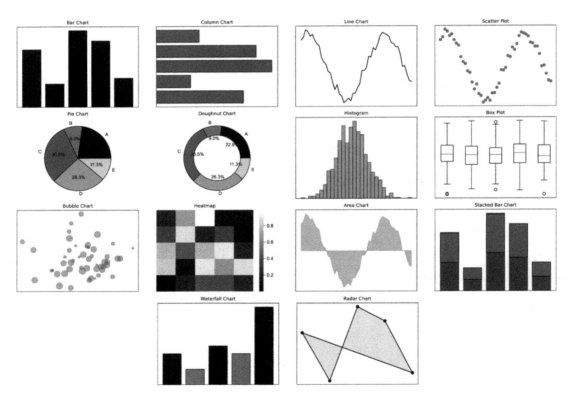

Figure 11-2. *An illustration of commonly used chart types*

As shown in Figure 11-2, visualization techniques serve distinct purposes in data storytelling. Bar charts are effective for comparing categorical data, such as sales figures across different product lines. Line charts are invaluable for illustrating trends over time, like website traffic fluctuations throughout the year. Scatterplots reveal relationships between two numerical variables, showcasing correlations or patterns. Histograms display the distribution of a single numerical variable, highlighting frequency and range. Maps are essential for geographical data, allowing spatial analysis and pattern recognition. Lastly, dashboards can combine multiple visualizations into a unified interface, providing a comprehensive overview of key performance indicators.

Visualization Tools Recommended for PostgreSQL

PostgreSQL, a robust relational database, can be seamlessly integrated with various visualization tools to bring data to life. Tableau and Power BI offer user-friendly interfaces for creating interactive dashboards, while Grafana is excellent for time-series data visualization, often used for monitoring. Python libraries such as Matplotlib and Seaborn provide a more programmatic approach, offering a wide range of customization options, as illustrated in Figure 11-2. To export data from PostgreSQL to these tools, you can perform standard SQL queries to extract the desired data and then export it as a CSV file, or use direct database connectors provided by the visualization tools. For example, Tableau and Power BI allow direct connections to PostgreSQL databases via ODBC or JDBC drivers. As a result, the visualizations can reflect the most current information, whether they are updated live or on a schedule.

Beyond the Presentation: How to Guide Your Audience

Effective data presentations do not end with the last slide. Ensure that audience members understand, retain, and act on insights by guiding them beyond the presentation. This involves providing supporting materials, encouraging further exploration, and fostering a data-driven culture.

Supporting Materials

To enhance understanding and credibility, it's essential to share resources that support the analysis. Transparency in data sourcing allows audiences to verify information and understand reliability. There are limitations in every dataset, as well as data challenges such as missing values or biases. Furthermore, creating a structured list of key terms, metrics, and definitions ensures consistency in interpretation, especially when working with complex datasets or technical terminology.

Encouraging Further Explorationt

A data presentation should inspire a sense of curiosity in the audience. Rather than simply delivering conclusions, presenters should engage the audience in a discussion of new questions. Discussing the questions behind SQL queries can lead to better insights or more insightful questions that can uncover better data findings. Also, providing interactive tools like shared dashboards or datasets enables hands-on exploration, empowering the audience to uncover insights independently.

Creating a Data-Driven Culture

Beyond a single presentation, the goal is to create an environment where data is at the core of decision-making. By using data to inform strategies, presenters can help audiences shift mindsets from intuition-based to evidence-based decision-making. By incorporating this strategy, presenters can move beyond just sharing insights—they can equip their audience with the tools, mindset, and motivation to use data effectively long after the presentation ends.

Final Thoughts: The Data Storyteller's Legacy in the Age of Artificial Intelligence (AI)

As you've learned, data storytelling is the ability to transform complex data into engaging, meaningful narratives. Data storytellers need clarity, context, and engagement. By clearly presenting data with context and relevance, storytellers can connect with their audience on an emotional level. This makes the data not just informative but also interesting by understanding the audience's needs. It also selects the right visualizations and emphasizes the story's core message.

Data storytelling is evolving as technology advances. Tools and techniques will evolve, but the core need to communicate data effectively will remain. This encourages analysts to continue refining their data storytelling skills, which is crucial. The future of data storytelling will be dynamic, but its goal will always be to make data accessible, relatable, and actionable for all.

Undoubtedly, the emergence of phenomena such as large language models (LLMs) and advanced language models has raised many concerns for those involved in this field or planning to enter it in the future. Generative AI and LLMs have revolutionized query writing and data analysis in recent years. These technologies enable automated SQL query generation, natural language processing for extracting insights, and even predictive analytics. By reducing technical expertise, they empower a broader audience to interact with data more simply. As these AI tools continue to advance, they will further enhance the speed, efficiency, and accuracy of data analysis. This will make it easier to uncover insights and create compelling narratives from large datasets.

Despite all the advantages that come with the emergence of AI for writing queries, a great question has also arisen. While LLMs can indeed generate SQL queries and help automate data analysis tasks, are there still logical reasons to learn SQL?

The answer that I give to this question as the author of this book when writing the current chapter in April 2025 is "most definitely yes," analysts still need to learn programming languages, query writing, and many of the skills that AI can do. In fact, I think that knowing SQL allows analysts to understand the logic behind queries. This helps them not only understand what a model generates but also modify and optimize queries according to specific needs. Without understanding SQL, they may struggle to adapt queries when they encounter complex data structures or specific requirements that the LLM doesn't automatically handle well. LLMs can generate queries based on patterns they've been trained on, but they might not always generate the most efficient or contextually appropriate query for each particular situation. Through the study of SQL, analysts learn how to customize queries to suit their exact needs, refine query performance, and ensure they follow best practices. LLM queries that don't return the expected results or perform poorly can be diagnosed with SQL knowledge. It is possible to interpret error messages, identify performance issues, and optimize queries.

In summary, while LLMs can assist in generating queries, SQL remains a fundamental skill that allows analysts to manage, understand, and troubleshoot their database systems.

Anyway, I hope your time and effort spent reading and practicing this book will help you learn data analysis, query writing, and the ability to tell stories in the world of data. In conclusion, I must say that reading just one book is not sufficient to become an expert data analyst. I suggest that you always update your knowledge and skills. As a data analyst, I recommend gaining a solid understanding of the basics and keeping yourself motivated to learn more. As you continue on your data storytelling journey, remember that every dataset has a story waiting to be told. The world of data storytelling is vast and ever-changing, but your role in making data meaningful remains valuable.

Summary

This chapter explored the art of effectively presenting a data story, ensuring insights are clear, compelling, and actionable. You learned how to structure your narrative, select impactful visualizations, and engage your audience with storytelling principles. A data-driven culture was also discussed, as well as strategies to guide audience interpretation, encourage further exploration, and do further research. Using these strategies, raw data can be transformed into meaningful insights that facilitate informed decisions.

Key Points

- Data stories need to be structured with clear narratives and compelling visualizations to ensure they are both accessible and engaging.

- Adapting presentations to different audiences enhances comprehension and encourages decision-making based on data.

- While AI and language models can assist in generating queries, SQL remains a fundamental skill which allows analysts to manage, understand, and troubleshoot their databases.

Key Takeaways

- **Effective data narratives:** The ability to present a data story that is clearly structured with a clear beginning, middle, and end increases audience engagement and comprehension.

- **Strategic visualization selection**: Choosing the right chart and visual elements will ensure that your data is communicated clearly and convincingly.

- **Audience-centered communication**: Making presentations relevant to different audiences promotes better understanding and facilitates data-driven decision-making.

APPENDIX A

SQL Syntax Reference Guide

This appendix is designed as a quick reference guide for readers working with SQL in data analysis. It focuses on core elements like essential SQL syntax, key SQL keywords, and standard command structures. Rather than diving into exhaustive theoretical details, the guide highlights practical syntax patterns commonly used in querying, manipulating, and managing data.

SQL Terminology

Whenever it comes to SQL, it's essential to distinguish between several core components that make up the language's structure and functionality. At the top level are statements, which are complete instructions that perform actions on a database. These instructions include SELECT to retrieve data, INSERT to add data, UPDATE to modify data, DELETE to remove data, and CREATE TABLE to define SQL's structures.

Within these statements are clauses, which act as building blocks that refine the statement's intent. Clauses like FROM, WHERE, GROUP BY, ORDER BY, and HAVING specify sources, filters, and grouping logic. Operations refer to specific actions applied to data, including arithmetic (+, -), comparison (=, <, >), and logical operations (AND, OR, NOT). SQL also includes functions, which are predefined routines for calculations and transformations—such as SUM(), AVG(), COUNT(), MAX(), and string and date functions like UPPER() or NOW(). These components often come together in expressions, which combine values, operators, and functions to produce a single result. The term *command* is sometimes used interchangeably with statement, especially in command-line interfaces where interactions with the database system occur. Table A-1 shows this terminology.

© Hamed Tabrizchi 2025
H. Tabrizchi, *Narrative SQL*, https://doi.org/10.1007/979-8-8688-1560-7

Table A-1. *SQL Terminology*

Component	Description	Examples
Statements	Complete instructions that perform actions on the database.	SELECT retrieves data
		INSERT adds new data
		UPDATE modifies existing data
		DELETE removes data
		CREATE TABLE defines a new table
Clauses	Parts of SQL statements that specify sources, conditions, or actions.	FROM specifies source table(s)
		WHERE filters rows based on conditions
		GROUP BY groups rows with similar values
		ORDER BY sorts the result set
		HAVING filters grouped results
Operations	Actions applied to values in expressions.	Arithmetic: +, -, *, /
		Comparison: =, <, >, !=
		Logical: AND, OR, and NOT
Functions	Predefined routines for calculations or data manipulation.	SUM( ) adds values
		AVG( ) calculates average
		COUNT( ) counts rows
		MAX( ), MIN( ) finds extremes
		UPPER( ), LOWER( ) changes case
		NOW( ), DATEADD( )

SQL Basics

SQL statements are categorized into functional groups based on their role in database management. Table A-2 illustrates the four main types of SQL statements, categorized by their purpose, description, and common commands used in database operations.

Table A-2. *Four Main Types of SQL Statements*

Category	Description	Common Commands
Data Definition Language (DDL)	Defines and modifies database schema and structure.	`CREATE, ALTER, DROP, TRUNCATE`
Data Manipulation Language (DML)	Manages data within tables.	`SELECT, INSERT, UPDATE, DELETE`
Data Control Language (DCL)	Controls access and permissions.	`GRANT, REVOKE`
Transaction Control Language (TCL)	Manages transactions within the database.	`BEGIN, COMMIT, ROLLBACK, SAVEPOINT`

Among all the mentioned main types of SQL statements, DML is the most frequently used for data analysis. Table A-3 shows four essential DML commands, highlighting their purpose, basic syntax, and example use cases for managing data within a database.

Table A-3. *Essential DML Commands*

Command	Purpose	Basic Syntax	Example Use Case
`SELECT`	Retrieves data from one or more tables.	`SELECT column1, column2 FROM table_name WHERE condition;`	Get all customer names from the `Customers` table.
`INSERT`	Adds new rows of data to a table.	`INSERT INTO table_name (column1, column2) VALUES (value1, value2);`	Add a new product to the `Products` table.
`UPDATE`	Modifies existing data in a table.	`UPDATE table_name SET column1 = value1 WHERE condition;`	Change a customer's email address in the `Customers` table.
`DELETE`	Removes rows from a table.	`DELETE FROM table_name WHERE condition;`	Remove discontinued products from the `Products` table.

Table A-4 illustrates key DDL commands, outlining their purpose, basic syntax, and example use cases for defining and modifying database structures.

Table A-4. *Essential DDL Commands*

Command	Purpose	Basic Syntax	Example Use Case
CREATE TABLE	Defines a new table and its structure.	CREATE TABLE table_name (column1 datatype, column2 datatype, ...);	Create a users table with ID, name, and email.
ALTER TABLE	Modifies an existing table.	ALTER TABLE table_name ADD column_name datatype; ALTER TABLE table_name DROP COLUMN column_name;	Add an Age column to the Users table.
DROP TABLE	Permanently removes a table and its data.	DROP TABLE table_name;	Delete the archived_orders table.
TRUNCATE TABLE	Removes all rows from a table (faster than DELETE).	TRUNCATE TABLE table_name;	Quickly clear all data from a Logs table.

SQL supports a range of data types to represent different kinds of values. Table A-5 summarizes commonly used, PostgreSQL-compatible, SQL data types.

Table A-5. *Common Data Types (PostgreSQL-Compatible)*

Data Type	Description	Example Use Case
INT, INTEGER	Whole numbers	Counting records, IDs
DECIMAL(p,s), NUMERIC(p,s)	Exact decimal numbers	Financial data, currency
REAL, DOUBLE PRECISION	Approximate numeric values	Scientific or statistical data
CHAR(n), VARCHAR(n)	Fixed or variable-length strings	Names, email addresses
TEXT	Large text strings	Comments, descriptions
DATE, TIME, TIMESTAMP	Date and time values	Birthdates, transaction logs
BOOLEAN	True/false values	Flags, binary states
BYTEA	Binary data (alternative to BLOB)	Storing images, files, documents

Identifiers and Naming Conventions

In SQL, identifiers refer to the names of database objects such as tables, columns, indexes, and functions. Naming these objects consistently and correctly ensures better code readability, fewer syntax errors, and improved maintainability. Table A-6 summarizes the essential rules and best practices related to identifiers in PostgreSQL.

Table A-6. *Essential Rules and Best Practices Related to Identifiers in PostgreSQL*

Category	Rule/Best Practice	Description	Examples
Basic structure	Start with a letter or underscore	Identifiers must begin with an alphabetic character (a-z, A-Z) or an underscore (_).	`users`, `product_details`, `_temp_data`
	Contain letters, digits, and underscores	Subsequent characters can be letters, digits (0-9), or underscores. Other characters are generally not allowed without quotation marks.	`user_id`, `order2023`, `item_count`
	Case sensitivity (Unquoted)	Unquoted identifiers are folded to lowercase. So, `TableName`, `tablename`, and `TABLENAME` are all interpreted as `tablename`.	Creating a table as `MyTable` will be referenced as `mytable`.
	Case sensitivity (Quoted)	Identifiers enclosed in double quotes (") are treated as case-sensitive. `"MyTable"` is different from `"mytable"`.	`SELECT "FirstName" FROM "Users";`
Reserved words	Avoid reserved SQL keywords (unquoted)	Do not use unquoted SQL keywords (e.g., `SELECT`, `FROM`, `WHERE`, `USER`) as identifiers. This will lead to syntax errors.	Avoid: `SELECT`, `user`, `group` (unquoted)
	Use double quotes to use reserved words	If you absolutely must use a reserved word as an identifier, enclose it in double quotes. This is generally discouraged.	`"SELECT"`, `"user"`, `"group"`

(continued)

Table A-6. (*continued*)

Category	Rule/Best Practice	Description	Examples
Length limits	Maximum length	PostgreSQL has a maximum length for identifiers, typically around 63 bytes. While you can often exceed this, it's best to keep them reasonable.	Consider shorter, more descriptive names
Best practices	Be descriptive and meaningful	Choose names that clearly indicate the purpose of the database object.	Instead of `t1`, use `customers`. Instead of `col1`, use `customer_name`.
	Maintain consistency	Adopt a consistent naming convention throughout your database schema (e.g., snake_case, camelCase).	`customer_order` (snake_case) vs. `customerOrder` (camelCase)
	Use underscores for separation	Underscores are commonly used to separate words within an identifier, improving readability.	`product_category`, `order_item_id`
	Avoid special characters (unquoted)	While technically some special characters might be allowed with quoting, it's best to avoid them in general for simplicity and portability.	Avoid: `data-value`, `user name` (unquoted)
	Consider plural vs. singular for tables	There's no strict rule, but be consistent. Plural is often used for tables representing collections of entities.	`products`, `orders`
	Use singular for column names representing single values	Column names usually represent a single attribute of a row.	`customer_id`, `product_name`, `order_date`

JOINs and Subqueries

Table A-7 shows various SQL JOIN types and subquery concepts, detailing their purpose, basic syntax, and how they are used to retrieve and relate data across tables.

Table A-7. *SQL JOIN Types and Subquery Concepts*

Concept	Purpose	Basic Syntax
INNER JOIN	Returns only rows with matching values in both tables.	`SELECT * FROM table1 INNER JOIN table2 ON table1.id = table2.id;`
LEFT JOIN	Returns all rows from the left table and matched rows from the right table.	`SELECT * FROM table1 LEFT JOIN table2 ON table1.id = table2.id;`
RIGHT JOIN	Returns all rows from the right table and matched rows from the left table.	`SELECT * FROM table1 RIGHT JOIN table2 ON table1.id = table2.id;`
FULL JOIN	Returns all rows when there is a match in either table.	`SELECT * FROM table1 FULL JOIN table2 ON table1.id = table2.id;`
CROSS JOIN	Returns the Cartesian product of both tables.	`SELECT * FROM table1 CROSS JOIN table2;`
SELF JOIN	Joins a table to itself using table aliases.	`SELECT a.name, b.name FROM employees a JOIN employees b ON a.manager_id = b.id;`
Non-correlated subquery	Executes independently of the outer query.	`SELECT name FROM employees WHERE dept_id IN (SELECT id FROM departments);`
Correlated subquery	References a column from the outer query inside the subquery.	`SELECT name FROM employees e WHERE salary > (SELECT AVG(salary) FROM employees WHERE dept_id = e.dept_id);`

Aggregate Functions and Grouping

Table A-8 illustrates key SQL aggregate functions and grouping clauses, explaining their purpose, syntax, and how they are used to summarize and filter data in queries.

Table A-8. *Key SQL Aggregate Functions and Grouping Clauses*

Concept	Purpose	Basic Syntax
COUNT()	Counts the number of rows or non-NULL values.	SELECT COUNT(*) FROM table;
SUM()	Calculates the total of a numeric column.	SELECT SUM(column) FROM table;
AVG()	Computes the average of a numeric column.	SELECT AVG(column) FROM table;
MIN()	Finds the smallest value in a column.	SELECT MIN(column) FROM table;
MAX()	Finds the largest value in a column.	SELECT MAX(column) FROM table;
GROUP BYclause	Groups rows based on one or more columns.	SELECT column, AGG_FUNC(column) FROM table GROUP BY column;
HAVING clause	Filters groups based on aggregate conditions.	SELECT column, AGG_FUNC(column) FROM table GROUP BY column HAVING condition;

Conditional Logic and Control Flow

Table A-9 shows SQL conditional expressions and functions, describing their purpose, syntax, and how they help in handling logic and NULL values within queries.

Table A-9. *SQL Conditional Expressions and Functions*

Concept	Purpose	Basic Syntax
CASE statement	Evaluates conditions and returns values based on those conditions.	`SELECT CASE WHEN condition THEN result ELSE alternative END FROM table;`
COALESCE()	Returns the first non-NULL value in a list of expressions.	`SELECT COALESCE(column1, column2, default_value) FROM table;`
NULLIF()	Returns NULL if two expressions are equal; otherwise, returns the first.	`SELECT NULLIF(expression1, expression2) FROM table;`
Conditional expressions	Used in WHERE, HAVING, and other clauses to filter or evaluate logic.	`SELECT * FROM table WHERE condition;`

Common Table Expressions (CTEs) and Views

Table A-10 illustrates the use of CTEs and views in SQL, including their purpose, syntax, and how they support query modularization and virtual table creation.

Table A-10. *Common Table Expressions and Views in SQL*

Concept	Purpose	Basic Syntax
WITH (non-recursive CTE)	Defines a temporary result set to be used in a larger query.	`WITH cte_name AS (SELECT ... FROM table) SELECT ... FROM cte_name;`
WITH RECURSIVE	Defines a recursive CTE to handle hierarchical or iterative data.	`WITH RECURSIVE cte_name AS (SELECT ... UNION ALL SELECT ... FROM cte_name ...) SELECT ... FROM cte_name;`
CREATE VIEW	Creates a virtual table based on a query result.	`CREATE VIEW view_name AS SELECT ... FROM table;`
Use/View Query	Queries data from the created view like a table.	`SELECT ... FROM view_name;`
DROP VIEW	Deletes a view from the database.	`DROP VIEW view_name;`

Indexing and Optimization

Table A-11 shows SQL indexing and performance optimization techniques, including index creation and analysis tools like EXPLAIN and ANALYZE that improve query efficiency.

Table A-11. *SQL Indexing and Performance Optimization Techniques*

Concept	Purpose	Basic Syntax
CREATE INDEX	Creates an index on one or more columns to speed up query performance.	CREATE INDEX index_name ON table (column);
DROP INDEX	Removes an existing index from the database.	DROP INDEX index_name;
Query performance tips	Improve efficiency using selective columns in WHERE clauses, avoid SELECT *, and normalize the data.	(Best practices; not syntax-based)
EXPLAIN	Shows the execution plan for an SQL query, helping analyze performance.	EXPLAIN SELECT ... FROM table WHERE condition;
ANALYZE	Executes the query and provides detailed performance statistics.	EXPLAIN ANALYZE SELECT ... FROM table WHERE condition;

Stored Procedures and Functions

Table A-12 illustrates the use of stored procedures and functions in SQL, including their creation, invocation, and example use cases for encapsulating tasks and returning values.

Table A-12. *The Use of Stored Procedures and Functions in SQL*

Concept	Purpose	Basic Syntax	Example
Create stored procedure	Encapsulates a sequence of SQL statements to perform a task.	`CREATE PROCEDURE procedure_name() LANGUAGE plpgsql AS $$ BEGIN --statements END; $$;`	`CREATE PROCEDURE log_action() LANGUAGE plpgsql AS $$ BEGIN INSERT INTO logs(action) VALUES ('test'); END; $$;`
Call stored procedure	Executes a previously defined stored procedure.	`CALL procedure_name();`	`CALL log_action();`
Create function	Returns a value and can be used in SQL expressions.	`CREATE FUNCTION function_name() RETURNS return_type LANGUAGE plpgsql AS $$ BEGIN RETURN value; END; $$;`	`CREATE FUNCTION add_numbers(a INT, b INT) RETURNS INT LANGUAGE plpgsql AS $$ BEGIN RETURN a + b; END; $$;`
Use function	Invokes the function within queries or statements.	`SELECT function_name();`	`SELECT add_numbers (3, 5);`

Table A-13 highlights the differences in SQL dialects across various database systems (MySQL, PostgreSQL, SQL Server, and Oracle), showcasing variations in features like auto-increment, string functions, NULL handling, and query syntax.

Table A-13. *SQL Dialect Differences*

Feature	MySQL	PostgreSQL	SQL Server	Oracle
Auto-increment	AUTO_ INCREMENT	SERIAL / BIGSERIAL	IDENTITY	SEQUENCE with NEXTVAL
Limit rows	LIMIT n	LIMIT n	TOP n	FETCH FIRST n ROWS ONLY (12c+)
NULL handling	IFNULL(expr, value)	COALESCE(expr, value)	ISNULL(expr, value)	NVL(expr, value)
String functions	LOWER(str), UPPER(str)	LOWER(str), UPPER(str)	LOWER(str), UPPER(str)	LOWER(str), UPPER(str)
String concatenation	CONCAT(str1, str2) or 'str		str2'	'str1
Date functions	CURDATE(), NOW()	CURRENT_DATE, CURRENT_ TIMESTAMP	GETDATE(), GETUTCDATE()	SYSDATE, CURRENT_DATE
Subqueries in SELECT	Supported	Supported	Supported	Supported
Boolean data type	TINYINT(1) or BOOLEAN (alias)	BOOLEAN	BIT	Use NUMBER(1) or CHAR(1) with constraints
Case sensitivity	Depends on OS	Case-sensitive	Case-insensitive by default	Case-insensitive by default
Join syntax	Standard SQL	Standard SQL	Standard SQL	Standard SQL

As shown in Table A-13, this table compares key SQL syntax and behavior differences across MySQL, PostgreSQL, SQL Server, and Oracle. For auto-incrementing columns, MySQL uses AUTO_INCREMENT, PostgreSQL uses SERIAL, SQL Server uses IDENTITY, and Oracle uses SEQUENCE. Row limiting is done using LIMIT in MySQL/PostgreSQL, and TOP in SQL Server. Each uses different functions for NULL handling and date operations. String functions like LOWER() and UPPER() are common across all. Subqueries are generally supported. Boolean types vary—PostgreSQL supports BOOLEAN, while SQL Server uses BIT. Case sensitivity and JOIN syntax are similar, although case behavior depends on the system or configuration.

PostgreSQL Cheat Sheet

This PostgreSQL Cheat Sheet provides a concise and practical reference to the most common SQL commands, functions, and features found in PostgreSQL databases. Table A-14 shows the basic SQL queries, covering a range of operations like selecting, filtering, sorting, and checking conditions, with examples of their syntax and use cases.

Table A-14. *Basic Queries*

Category	Type	Basic Syntax
SELECT	Basic SELECT	`SELECT column1, column2 FROM table_name;`
	Select all columns	`SELECT * FROM table_name;`
	Select with alias	`SELECT column1 AS col1, column2 AS "Column 2" FROM table_name;`
	Select distinct values	`SELECT DISTINCT column1 FROM table_name;`
	Select with condition	`SELECT * FROM table_name WHERE condition;`
	Limit results	`SELECT * FROM table_name LIMIT 10;`
	Offset results	`SELECT * FROM table_name OFFSET 10 LIMIT 10;`

(continued)

Table A-14. (*continued*)

Category	Type	Basic Syntax
Filtering	Basic comparison operators	`SELECT * FROM table_name WHERE column1 = value;` -- Equal `SELECT * FROMtable_name WHERE column1 > value;` -- Greater than `SELECT * FROM table_name WHERE column1 >= value;` -- Greater than or equal `SELECT * FROM table_name WHERE column1 < value;` -- Less than `SELECT * FROMtable_name WHERE column1 <= value;` -- Less than or equal `SELECT * FROM table_name WHERE column1 != value;` -- Not equal `SELECT * FROM table_name WHERE column1 <> value;` -- Not equal (alternative)
	Logical operators	`SELECT * FROM table_name WHERE condition1 AND condition2;` -- AND `SELECT * FROMtable_name WHERE condition1 OR condition2;` -- OR `SELECT * FROM table_name WHERE NOT condition;` -- NOT
	NULL checks	`SELECT * FROM table_name WHERE column1 IS NULL;` `SELECT * FROM table_name WHEREcolumn1 IS NOT NULL;`
	Pattern matching	`SELECT * FROM table_name WHERE column1 LIKE 'pattern%';` -- Starts with 'pattern' `SELECT *FROM table_name WHERE column1 LIKE '%pattern';` -- Ends with 'pattern' `SELECT * FROM table_name WHEREcolumn1 LIKE '%pattern%';` -- Contains 'pattern' `SELECT * FROM table_name WHERE column1 ILIKE '%pattern%';` -- Case-insensitive LIKE

(*continued*)

Table A-14. (*continued*)

Category	Type	Basic Syntax
	Range checks	`SELECT * FROM table_name WHERE column1 BETWEEN value1 AND value2;` `SELECT * FROM table_name WHERE column1 NOT BETWEEN value1 AND value2;`
	List checks	`SELECT * FROM table_name WHERE column1 IN (value1, value2, value3);` `SELECT * FROM table_name WHERE column1 NOT IN (value1, value2, value3);`

Sorting	Basic sorting	`SELECT * FROM table_name ORDER BY column1;`
		`-- ASC by default`
		`SELECT * FROM table_name ORDER BY column1 ASC;`
		`-- Ascending`
		`SELECT * FROM table_name ORDER BY column1 DESC;`
		`-- Descending`
		`SELECT * FROM table_name ORDER BY column1 DESC, column2 ASC;   -- Multiple columns`

Table A-15 illustrates SQL aggregation and grouping functions, showcasing how to perform operations like counting, summing, and averaging data, as well as how to group and filter aggregated results.

Table A-15. *Aggregations and Grouping*

Category	Type	Basic Syntax

Aggregate functions	Basic aggregates	`SELECT COUNT(*) FROM table_name;`
		 `-- Count rows`
		 `SELECT COUNT(column1) FROM table_name;`
		 `-- Count non-null values`
		 `SELECT COUNT(DISTINCT column1) FROM table_name;` `-- Count distinct values`
		 `SELECT SUM(column1) FROM table_name;`
		 `-- Sum`
		 `SELECT AVG(column1) FROM table_name;`
		 `-- Average`
		 `SELECT MIN(column1) FROM table_name;`
		 `-- Minimum`
		 `SELECT MAX(column1) FROM table_name;`
		 `-- Maximum`
		 `SELECT STDDEV(column1) FROM table_name;`
		 `-- Standard deviation`
		 `SELECT VARIANCE(column1) FROM table_name;`
		 `-- Variance`

GROUP BY	Basic grouping	`SELECT column1, COUNT(*)` `FROM table_name GROUP BY column1;`

	Multiple columns	`SELECT column1, column2, SUM(column3)`
		 `FROM table_name`
		 `GROUP BY column1, column2;`

	With HAVING (filters on aggregated data)	`SELECT column1, COUNT(*)`
		 `FROM table_name`
		 `GROUP BY column1`
		 `HAVING COUNT(*) > 5;`

Table A-16 demonstrates different SQL join types, including INNER JOIN, LEFT JOIN, RIGHT JOIN, FULL OUTER JOIN, CROSS JOIN, and SELF JOIN, along with their basic syntax and use cases for combining data from multiple tables.

Table A-16. *Joins*

Category	Type	Basic Syntax
Basic JOINs	INNER JOIN	SELECT t1.column1, t2.column2 FROM table1 t1 INNER JOIN table2 t2 ON t1.id = t2.id;
	LEFT JOIN	SELECT t1.column1, t2.column2 FROM table1 t1 LEFT JOIN table2 t2 ON t1.id = t2.id;
	RIGHT JOIN	SELECT t1.column1, t2.column2 FROM table1 t1 RIGHT JOIN table2 t2 ON t1.id = t2.id;
	FULL OUTER JOIN	SELECT t1.column1, t2.column2 FROM table1 t1 FULL OUTER JOIN table2 t2 ON t1.id = t2.id;
	CROSS JOIN	SELECT t1.column1, t2.column2 FROM table1 t1 CROSS JOIN table2 t2;
Self JOIN	Self-join (joining a table to itself)	SELECT a.column1, b.column2 FROM table_name a JOIN table_name b ON a.id = b.parent_id;

Table A-17 illustrates different types of subqueries in SQL, including basic subqueries in the WHERE and FROM clauses, correlated subqueries, and CTEs, with syntax for both basic and recursive CTEs.

Table A-17. *Subqueries*

Category	Type	Basic Syntax

Basic subqueries	Subquery in WHERE	`SELECT * FROM table1`
		`WHERE column1 IN (SELECT column1 FROM`
		`table2);`

	-- Subquery in FROM	`SELECT a.column1, b.column2`
		`FROM table1 a, (SELECT * FROM table2) b`
		`WHERE a.id = b.id;`

	Correlated subquery	`SELECT *`
		`FROM table1 t1`
		`WHERE column1 > (`
		`SELECT AVG(column1)`
		`FROM table1 t2`
		`WHERE t2.category = t1.category`
		`);`

(*continued*)

Table A-17. (*continued*)

Category	Type	Basic Syntax
Common Table Expressions (CTEs)	Basic CTE	WITH cte_name AS (SELECT column1, column2 FROM table_name WHERE condition) SELECT * FROM cte_name;
	Multiple CTEs	WITH cte1 AS (SELECT * FROM table1 WHERE condition1), cte2 AS (SELECT * FROM table2 WHERE condition2) SELECT * FROM cte1 JOIN cte2 ON cte1.id = cte2.id;
	Recursive CTE	WITH RECURSIVE cte_name AS (-- Base case SELECT * FROM table_name WHERE condition UNION ALL -- Recursive case SELECT t.* FROM table_name t JOIN cte_name c ON c.id = t.parent_id) SELECT * FROM cte_name;

Table A-18 provides basic SQL commands for data manipulation, including INSERT for adding data, UPDATE for modifying records, and DELETE for removing data. It covers syntax for basic operations like inserting multiple rows, updating multiple columns, and deleting with a subquery.

Table A-18. *Data Modification*

Category	Type	Basic Syntax
INSERT	Basic Insert	INSERT INTO table_name (column1, column2) VALUES (value1, value2);
	Multiple rows	INSERT INTO table_name (column1, column2) VALUES (value1, value2), (value3, value4);
	Insert from SELECT	INSERT INTO table_name (column1, column2) SELECT column1, column2 FROM source_table WHERE condition;
UPDATE	Basic Update	UPDATE table_name SET column1 = value1 WHERE condition;
	Update multiple columns	UPDATE table_name SET column1 = value1, column2 = value2 WHERE condition;
	Update with JOIN	UPDATE table1 t1 SET t1.column1 = t2.column1 FROM table2 t2 WHERE t1.id = t2.id;
DELETE	Basic Delete	DELETE FROM table_name WHERE condition;
	Delete all rows	DELETE FROM table_name;
	Delete with subquery	DELETE FROM table_name WHERE column1 IN (SELECT column1 FROM other_table WHERE condition);

Table A-19 describes SQL window functions, including ROW_NUMBER() and PARTITION BY, as well as common window functions such as SUM(), AVG(), MIN(), MAX(), and COUNT(). It also covers ranking functions like RANK(), DENSE_RANK(), NTILE(), as well as LEAD() and LAG() for accessing data in previous or next rows. These functions are used for advanced data analysis and calculations across partitions or ordered result sets.

Table A-19. *Window Functions*

Category	Basic Syntax
Row number	SELECT column1, ROW_NUMBER() OVER (ORDER BY column2) AS row_num FROM table_name;
Partition by	SELECT column1, column2, ROW_NUMBER() OVER (PARTITION BY column1 ORDER BY column2) AS row_num FROM table_name;
Common window functions	SELECT column1, column2, SUM(column2) OVER (PARTITION BY column1) AS sum, AVG(column2) OVER (PARTITION BY column1) AS avg, MIN(column2) OVER (PARTITION BY column1) AS min, MAX(column2) OVER (PARTITION BY column1) AS max, COUNT(*) OVER (PARTITION BY column1) AS count FROM table_name;

(*continued*)

Table A-19. (*continued*)

Category	Basic Syntax
Ranking functions	```SELECT column1, column2, ROW_NUMBER() OVER (ORDER BY column2) AS row_num, RANK() OVER (ORDER BY column2) AS rank, DENSE_RANK() OVER (ORDER BY column2) AS dense_rank, NTILE(4) OVER (ORDER BY column2) AS quartile FROM table_name;```
Lead and lag functions	```SELECT column1, column2, LEAD(column2) OVER (ORDER BY column1) AS next_value, LAG(column2) OVER (ORDER BY column1) AS prev_value FROM table_name;```

Table A-20 provides SQL syntax for working with date and time, covering operations like retrieving the current date and time, performing date arithmetic, extracting specific components of dates, and formatting date and time values.

Table A-20. *Date and Time Operations*

Category	Basic Syntax

Current date and time	```SELECT CURRENT_DATE; -- Current date```
	```SELECT CURRENT_TIME; -- Current time```
	```SELECT CURRENT_TIMESTAMP; -- Current date and time```
	```SELECT NOW(); -- Current timestamp```

Date/time operations	```SELECT date_column + INTERVAL '1 day';```
	```-- Add 1 day```
	```SELECT date_column + INTERVAL '2 hours';```
	```-- Add 2 hours```
	```SELECT date_column + INTERVAL '3 months';```
	```-- Add 3 months```
	```SELECT date_column - INTERVAL '1 year';```
	```-- Subtract 1 year```

Date/time extraction	```SELECT EXTRACT(YEAR FROM date_column);```
	```-- Get year```
	```SELECT EXTRACT(MONTH FROM date_column);```
	```-- Get month```
	```SELECT EXTRACT(DAY FROM date_column);```
	```-- Get day```
	```SELECT EXTRACT(HOUR FROM timestamp_column);```
	```-- Get hour```
	```SELECT EXTRACT(DOW FROM date_column);```
	```-- Day of week (0-6)```

Date/time formatting	```SELECT TO_CHAR(date_column, 'YYYY-MM-DD');```
	```-- Format date```
	```SELECT TO_CHAR(timestamp_column, 'HH24:MI:SS');```
	```-- Format time```

Table A-21 covers SQL string operations, such as concatenating strings, extracting substrings, calculating string length, replacing text, and finding the position of a substring within a string.

Table A-21. *String Operations*

Category	Basic Syntax

String operations	```SELECT CONCAT(column1, ' ', column2);```
	-- Concatenate
	```SELECT column1 || ' ' || column2;```
	-- Concat operator
	```SELECT UPPER(column1);```
	-- Uppercase
	```SELECT LOWER(column1);```
	-- Lowercase
	```SELECT INITCAP(column1);```
	-- Capitalize first letter
	```SELECT TRIM(' text ');```
	-- Trim spaces
	```SELECT LTRIM(' text');```
	-- Left trim
	```SELECT RTRIM('text ');```
	-- Right trim
	```SELECT SUBSTRING(column1, 1, 5);```
	-- Substring
	```SELECT LEFT(column1, 5);```
	-- Left n chars
	```SELECT RIGHT(column1, 5);```
	-- Right n chars
	```SELECT LENGTH(column1);```
	-- String length
	```SELECT REPLACE(column1, 'old', 'new');```
	-- Replace text
	```SELECT POSITION('needle' IN column1);```
	-- Find position

Table A-22 explains SQL conditional expressions, including the CASE expression for conditional logic, NULLIF for returning NULL when two values are equal, COALESCE for selecting the first non-NULL value, and GREATEST/LEAST for finding the highest or lowest value among multiple columns.

Table A-22. *Conditional Expressions*

Category	Type	Basic Syntax
Conditional Expressions	CASE expression	SELECT column1, CASE WHEN condition1 THEN result1 WHEN condition2 THEN result2 ELSE result3 END AS new_column FROM table_name;
	NULLIF (returns NULL if equal)	SELECT NULLIF(column1, 0) FROM table_name;
	COALESCE (returns first non-NULL value)	SELECT COALESCE(column1, column2, 'default') FROM table_name;
	GREATEST/LEAST	SELECT GREATEST(column1, column2, column3) FROM table_name; SELECT LEAST(column1, column2, column3) FROM table_name;

Table A-23 introduces SQL views, including how to create, manage, and use both regular and materialized views, with syntax for refreshing, altering, querying, and dropping them, as well as retrieving view metadata and performing updates when allowed.

Table A-23. *Views*

Category	Type	Basic Syntax

Creating views	Basic view creation	`CREATE VIEW view_name AS`
		`SELECT column1, column2`
		`FROM table_name`
		`WHERE condition;`

	Create or replace existing view	`CREATE OR REPLACE VIEW view_name AS`
		`SELECT column1, column2`
		`FROM table_name`
		`WHERE condition;`

	Materialized view (stored physically, needs refresh)	`CREATE MATERIALIZED VIEW mat_view_name AS`
		`SELECT column1, column2`
		`FROM table_name`
		`WHERE condition;`

	With options	`CREATE VIEW view_name WITH (security_`
		`barrier=true) AS`
		`SELECT column1, column2`
		`FROM table_name`
		`WHERE condition;`

Managing views	Refresh materialized view (update data)	`REFRESH MATERIALIZED VIEW mat_view_name;`
	Concurrent refresh (doesn't block queries)	`REFRESH MATERIALIZED VIEW CONCURRENTLY` `mat_view_name;`

	Alter view	`ALTER VIEW view_name RENAME TO new_name;`
		`ALTER VIEW view_name SET SCHEMA new_schema;`
		`ALTER VIEW view_name ALTER COLUMN column_`
		`name SET DEFAULT expression;`

	Drop view	`DROP VIEW view_name;`
		`DROP VIEW IF EXISTS view_name;`
		`DROP VIEW view_name CASCADE;  -- Also`
		`drops dependent objects`

	Drop materialized view	`DROP MATERIALIZED VIEW mat_view_name;`

(continued)

Table A-23. (*continued*)

Category	Type	Basic Syntax
View information	List all views in current database	`SELECT * FROM pg_views WHERE schemaname = 'public';`
	List all materialized views	`SELECT * FROM pg_matviews WHERE schemaname = 'public';`
	Get view definition	`SELECT pg_get_viewdef('view_name', true);`
Using views	Query a view like a regular table	`SELECT * FROM view_name WHERE column1 = value;`

	Join with a view	`SELECT t.column1, v.column2`
		`FROM table_name t`
		`JOIN view_name v ON t.id = v.id;`

	Update through a view (if possible)	`UPDATE view_name SET column1 = value WHERE condition;`

Table A-24 covers SQL indexing, detailing how to create, manage, and monitor different types of indexes to optimize queries, along with syntax for rebuilding, dropping, and checking index usage and size.

Table A-24. *Indexes*

Category	Type	Basic Syntax
Creating indexes	Basic index	CREATE INDEX index_name ON table_name (column_name);
	Multi-column index	CREATE INDEX index_name ON table_name (column1, column2);
	Unique index	CREATE UNIQUE INDEX index_name ON table_name (column_name);
	Partial index (only index certain rows)	CREATE INDEX index_name ON table_name (column_name) WHERE condition;
	Expression index	CREATE INDEX index_name ON table_name (lower(column_name));
	Index with specific method	CREATE INDEX index_name ON table_name USING BTREE (column_name); CREATE INDEX index_name ON table_name USING HASH (column_name); CREATE INDEX index_name ON table_name USING GIN (column_name); CREATE INDEX index_name ON table_name USING GIST (column_name);
	Create concurrently (doesn't block writes)	CREATE INDEX CONCURRENTLY index_name ON table_name (column_name);

(continued)

Table A-24. (*continued*)

Category	Type	Basic Syntax
Index types	B-tree (default): Good for equality and range queries	`CREATE INDEX index_name ON table_name USING BTREE (column_name);`
	Hash: Good only for equality comparisons	`CREATE INDEX index_name ON table_name USING HASH (column_name);`
	GiST: Good for geometrical data and full-text search	`CREATE INDEX index_name ON table_name USING GIST (geom_column);`
	GIN: Good for arrays and full-text search	`CREATE INDEX index_name ON table_name USING GIN (array_column);`
	BRIN (Block range index): Good for large tables with natural ordering	`CREATE INDEX index_name ON table_name USING BRIN (timestamp_column);`
Managing indexes	Rebuild index	`REINDEX INDEX index_name;`
	Rebuild all indexes on a table	`REINDEX TABLE table_name;`
	Drop index	`DROP INDEX index_name;` `DROP INDEX IF EXISTS index_name;` `DROP INDEX CONCURRENTLY index_name;`
	Disable/enable index	`ALTER INDEX index_name SET (fastupdate = off);`

(*continued*)

Table A-24. (*continued*)

Category	Type	Basic Syntax
Index information	List all indexes in database	SELECT indexname, tablename, indexdef FROM pg_indexes WHERE schemaname = 'public' ORDER BY tablename, indexname;
	List indexes on specific table	SELECT indexname, indexdef FROM pg_indexes WHERE tablename = 'table_name';
	Check index size	SELECT pg_size_pretty(pg_relation_ size(indexname::text)) as index_size FROM pg_indexes WHERE tablename = 'table_name';
	Check index usage	SELECT idstat.relname AS table_name, index.relname AS index_name, idx_scan AS times_used, idx_tup_read AS tuples_read, idx_tup_fetch AS tuples_fetched FROM pg_stat_user_indexes idxstat JOIN pg_stat_user_tables tabstat ON idxstat.relname = tabstat.relname WHERE idxstat.relname = 'table_name' ORDER BY idx_scan DESC;

This appendix has served as a compact SQL reference guide, covering core syntax, common commands, and differences.

APPENDIX B

Glossary of Terms

This appendix provides a glossary of terms commonly used in relation to SQL and data analysis, organized alphabetically. The glossary contains over 100 terms, providing a thorough reference for anyone working with PostgreSQL for data analysis purposes.

A

ACID (Atomicity, Consistency, Isolation, Durability): Set of properties that guarantee database transactions are processed reliably.

Aggregate functions: Functions that perform calculations across multiple rows and return a single value.

ANALYZE: Command that collects statistics about the contents of tables, used by the query planner to generate efficient execution plans.

Array: A data type that can store multiple values of the same type in a single column.

B

B-tree index: Default index type in PostgreSQL; a balanced tree structure optimized for range queries and equality conditions.

Backup: Copy of database data that can be used to restore the database in case of data loss.

BETWEEN: Operator used to test if a value falls within a specified range.

BIGINT: Integer data type capable of storing numbers between -9,223,372,036,854,775,808 and 9,223,372,036,854,775,807.

BOOLEAN: Data type that can store true, false, or NULL values.

BRIN (Block Range INdex): Index type suited for very large tables with natural ordering; stores summary information for block ranges.

© Hamed Tabrizchi 2025
H. Tabrizchi, *Narrative SQL*, https://doi.org/10.1007/979-8-8688-1560-7

C

Cast: Operation that converts a value from one data type to another.

CTE (Common Table Expression): Named temporary result set defined within a query that can be referenced multiple times.

COALESCE: Function that returns the first non-NULL value in a list.

Collation: Set of rules that determine how text data is sorted and compared in a database.

Composite type: User-defined data type consisting of multiple fields.

Concurrency control: Mechanisms that ensure data consistency when multiple transactions are executing simultaneously.

Connection pooling: Technique of maintaining a pool of database connections for reuse, improving performance.

Constraints: Rules enforced on data columns to maintain data integrity.

COPY: Command used for bulk loading or unloading data between PostgreSQL tables and files.

Correlation: Statistical measure that indicates the extent to which two variables fluctuate together.

Correlated subquery: Subquery that references columns from the outer query.

CUBE: Extension to GROUP BY that generates multiple grouping sets for multi-dimensional aggregation.

Cursor: Database object used to traverse through the result set of a query.

D

Data type: Attribute that specifies the type of data that an object can hold.

Database cluster: Collection of databases managed by a single PostgreSQL server instance.

Deadlock: Situation where two or more transactions are waiting for each other to release locks.

DISTINCT: Clause used to eliminate duplicate rows from a result set.

Domain: User-defined data type with constraints.

E

ENUM: Data type that represents a static, ordered set of values.

EXPLAIN: Command that shows the execution plan for a statement without actually executing it.

EXPLAIN ANALYZE: Command that shows the execution plan and actual runtime statistics.

Extension: Module that adds functionality to PostgreSQL, such as PostGIS.

Extract, Transform, Load (ETL): Process of copying data from source systems into a database.

F

Foreign key: Constraint that maintains referential integrity between two tables.

Full text search: Feature that allows searching through text data for specific words or phrases.

Function: Reusable block of SQL code that performs a specific task and can return a value.

G

Generalized Inverted Index (GIN): Index structure for handling cases where multiple values are stored in a single column, often used for arrays and full-text search.

Generalized Search Tree (GiST): Index structure supporting arbitrary indexing schemes, used for geometric data types and full-text search.

GROUP BY: Clause used to group rows with the same values into summary rows.

H

Hash index: Index structure that works well for equality comparisons but not for range queries.

HAVING: Clause used to filter groups in a GROUP BY query.

Histogram: Data structure that represents the distribution of values in a column.

HyperLogLog: Algorithm used for approximate distinct count calculations.

I

Index: Data structure that improves the speed of data retrieval operations.

Index-only scan: Query execution where all required data can be retrieved from the index without accessing the table.

INNER JOIN: Returns rows when there is a match in both tables.

Isolation level: Setting that determines how transaction integrity is enforced.

J

JOIN: SQL operation that combines rows from two or more tables based on a related column.

JSON/JSONB: Data types that store JSON (JavaScript Object Notation) data with different performance characteristics.

K

Key: Field or combination of fields used to identify a record or establish relationships between tables.

L

LATERAL JOIN: Allows subqueries in the FROM clause to reference columns from preceding items in the FROM clause.

LEFT JOIN: Returns all rows from the left table and matching rows from the right table.

LIMIT: Clause that restricts the number of rows returned by a query.

LISTEN/NOTIFY: Asynchronous notification system in PostgreSQL.

Locale: Set of parameters that defines language and cultural preferences.

Lock: Mechanism used to control concurrent access to data.

M

MADlib: An open-source library for scalable in-database analytics.

Materialized view: View that stores the result set physically and must be explicitly refreshed.

Multi-Version Concurrency Control (MVCC): Technique PostgreSQL uses to allow concurrent access to the database.

N

Natural JOIN: JOIN that automatically joins tables based on columns with the same name.

Normalization: Process of organizing data to reduce redundancy and improve data integrity.

NULL: Special marker used to indicate that a data value does not exist.

NULLIF: Function that returns NULL if two specified expressions are equal.

O

Object Identifier (OID): System column used internally by PostgreSQL.

OFFSET: Clause used to skip a specified number of rows before returning results.

Online Analytical Processing (OLAP): Data processing that enables users to analyze multidimensional data from multiple perspectives.

Online Transaction Processing (OLTP): Data processing focused on transaction-oriented applications.

Operator: Symbol that specifies an operation to be performed.

Optimizer: Component that determines the most efficient way to execute a query.

ORDER BY: Clause used to sort the result set.

P

Partition: Division of a large table into smaller, more manageable pieces.

Percentile: Value below which a given percentage of observations fall.

Prepared statement: SQL statement template that can be executed multiple times with different parameters.

Primary key: Column or group of columns that uniquely identifies each row in a table.

Procedural language: Language used to write stored procedures and functions (e.g., PL/pgSQL).

Q

Query: Request for data or information from a database.

Query plan: Series of steps PostgreSQL uses to access the data.

Query optimizer: Component that determines the most efficient way to execute a query.

R

RAISE: Statement used to report messages and raise errors in PL/pgSQL.

Recursive query: Query that references itself.

Referential integrity: Property that ensures relationships between tables remain consistent.

RETURNING: Clause that returns data from rows affected by INSERT, UPDATE, or DELETE.

ROLLUP: Extension to GROUP BY that creates hierarchical grouping sets.

Row-level security: Feature that restricts which rows users can access.

S

Schema: Namespace that contains named database objects.

Sequence: Object that generates unique numeric identifiers.

SERIAL: Pseudo-type for creating auto-incrementing integer columns.

Space-partitioned GiST (SP-GiST): Index for partitioned search trees, useful for data that can be divided into non-overlapping regions.

Statistical functions: Functions used for statistical analysis, including correlation, regression.

Subquery: Query nested inside another query.

T

Table: Collection of related data held in a structured format.

Table partitioning: Dividing a large table into smaller, more manageable pieces.

Temporary table: Table that exists only for the duration of a session or transaction.

Text search: Capability to search through text data using language-specific operations.

Transaction: Unit of work performed within a database management system.

Trigger: Function automatically executed in response to certain events on a table.

TRUNCATE: Command that removes all rows from a table more quickly than DELETE.

Tuple: Another term for a row in a relational database.

U

UNION/UNION ALL: Set operations that combine results of two or more SELECT statements.

UNIQUE **constraint**: Ensures all values in a column are different.

UPDATE: Statement used to modify existing records in a table.

UPSERT: Operation that either inserts a new row or updates an existing one (INSERT ... ON CONFLICT).

User-Defined Function (UDF): Custom function created by a database user.

UUID: Universally Unique Identifier, a 128-bit number used as an identifier.

V

VACUUM: Process that reclaims storage occupied by dead tuples.

View: Virtual table based on the result of an SQL statement.

Volatile function: Function whose result can change even with the same arguments.

W

WHERE: Clause used to filter records.

Window function: Function that performs a calculation across a set of table rows related to the current row.

WITH clause: Introduces CTEs to simplify complex queries.

Write-Ahead Logging (WAL): Method to ensure data integrity by writing changes to a log before they are applied.

X

XML: Data type and functions for working with XML data.

XPath: Query language for selecting nodes from an XML document.

Z

Z-order curve: Space-filling curve sometimes used in multi-dimensional indexing strategies.

Zero-based indexing: Array elements and string characters are numbered starting from 0 in PostgreSQL.

APPENDIX C

PostgreSQL Elements Reference

This reference provides a comprehensive alphabetical list of SQL statements, clauses, operations, and functions available in PostgreSQL.

A

ABS( )—Returns the absolute value of a number

ACOS( )—Returns the arc cosine of a number

ADD COLUMN—Adds a column to a table

ALL—Returns true if all subquery values meet the condition

ALTER DATABASE—Changes database attributes

ALTER SEQUENCE—Changes sequence generator properties

ALTER TABLE—Modifies a table definition

ALTER VIEW—Changes view attributes

AND—Logical operator that returns true when both conditions are true

ANY—Returns true if any subquery values meet the condition

ARRAY—Creates an array value

ARRAY_AGG( )—Aggregates values into an array

ARRAY_APPEND( )—Appends an element to the end of an array

ARRAY_CAT( )—Concatenates two arrays

ARRAY_LENGTH( )—Returns the length of an array dimension

AS—Renames a column or table with an alias

ASC—Sorts results in ascending order (the default)

ASIN( )—Returns the arc sine of a number

© Hamed Tabrizchi 2025
H. Tabrizchi, *Narrative SQL*, https://doi.org/10.1007/979-8-8688-1560-7

ATAN()—Returns the arc tangent of a number

ATAN2()—Returns the arc tangent of two numbers

AVG()—Returns the average value of a numeric column

B

BEGIN—Starts a transaction block

BETWEEN—Checks if a value is within a range

BIGINT—Integer data type with large range

BIGSERIAL—Auto-incrementing integer data type with large range

BIT_AND()—Bitwise AND aggregation

BIT_LENGTH()—Returns the number of bits in a string

BIT_OR()—Bitwise OR aggregation

BOOLEAN—Logical Boolean data type (true/false)

BTRIM()—Removes characters from both ends of a string

C

CASE—Provides conditional logic in SQL

CAST—Converts a value from one data type to another

CBRT()—Returns the cube root of a number

CEIL()—Rounds a number up to the nearest integer

CEILING()—Rounds a number up to the nearest integer

CHAR_LENGTH()—Returns the number of characters in a string

CHARACTER VARYING—Variable-length character data type

CHECK—Creates a constraint to check a condition

CHR()—Returns the character for the given Unicode code point

CLOSE—Closes a cursor

CLUSTER—Physically reorders a table based on an index

COALESCE()—Returns the first non-NULL value in a list

COLLATE—Specifies the collation for a sort operation

COLLATION FOR—Returns the collation of an expression

COLUMN—Specifies a column in a table

COMMIT—Commits the current transaction

CONCAT()—Concatenates strings together

CONCAT_WS()—Concatenates strings with a separator

CONSTRAINT—Defines a constraint on a table

CORR()—Returns the correlation coefficient

COS()—Returns the cosine of a number

COSH()—Returns the hyperbolic cosine of a number

COUNT()—Counts rows or non-NULL values

COVAR_POP()—Returns the population covariance

COVAR_SAMP()—Returns the sample covariance

CREATE DATABASE—Creates a new database

CREATE EXTENSION—Installs an extension

CREATE FUNCTION—Defines a new function

CREATE INDEX—Creates an index

CREATE MATERIALIZED VIEW—Creates a materialized view

CREATE RULE—Defines a new rewrite rule

CREATE SCHEMA—Creates a new schema

CREATE SEQUENCE—Creates a sequence generator

CREATE TABLE—Creates a new table

CREATE TABLE AS—Creates a table from a query result

CREATE TRIGGER—Creates a trigger

CREATE TYPE—Defines a new data type

CREATE USER—Creates a new database user

CREATE VIEW—Creates a view

CROSS JOIN—Produces the Cartesian product of two tables

CURRENT_DATE—Returns the current date

CURRENT_TIME—Returns the current time

CURRENT_TIMESTAMP—Returns the current date and time

CURRENT_USER—Returns the current user name

D

DATE—Date data type

DATE_PART()—Extracts a part of a date/time value

DATE_TRUNC()—Truncates a date/time value to specified precision

DECIMAL—Exact numeric data type with fixed precision and scale

DECLARE—Declares a cursor

DEFAULT—Specifies a default value for a column

DELETE—Deletes rows from a table

DESC—Sorts results in descending order

DISTINCT—Eliminates duplicate rows

DIV()—Integer division

DROP DATABASE—Removes a database

DROP FUNCTION—Removes a function

DROP INDEX—Removes an index

DROP TABLE—Removes a table

DROP TRIGGER—Removes a trigger

DROP VIEW—Removes a view

E

ELSE—Alternative condition in a CASE statement

END—Ends a conditional block or transaction

EXCEPT—Returns rows from the first query that are not in the second query

EXISTS—Tests for the existence of records in a subquery

EXP()—Returns e raised to a specified power

EXPLAIN—Shows the execution plan of a statement

EXTRACT()—Extracts a portion of a date or time value

F

FALSE—Logical false value

FETCH—Retrieves rows from a cursor

FIRST_VALUE()—Window function that returns the first value in an ordered set

FLOAT—Approximate numeric data type

FLOOR()—Rounds a number down to the nearest integer

FOR—Specifies a loop in PL/pgSQL

FOREIGN KEY—Defines a foreign key constraint

FROM—Specifies the tables to query

G

GENERATE_SERIES()—Generates a series of values

GREATEST()—Returns the largest value from a list

GROUP BY—Groups rows that have the same values

GROUPING SETS—Specifies multiple grouping sets in a single query

H

HAVING—Filters groups in a GROUP BY query

HOUR—Extracts the hour from a time or timestamp

I

IF—Conditional statement in PL/pgSQL

ILIKE—Case-insensitive LIKE

IN—Tests if a value matches any value in a list or subquery

INDEX—Creates an index on columns of a table

INITCAP()—Converts the first letter of each word to uppercase

INNER JOIN—Returns rows when there is a match in both tables

INSERT—Adds new rows to a table

INT or INTEGER—Integer data type

INTERSECT—Returns rows that are in both result sets

INTERVAL—Time interval data type

INTO—Specifies a destination for retrieved data

IS—Tests a value against Boolean truth values or NULL

IS NOT NULL—Tests if a value is not NULL

IS NULL—Tests if a value is NULL

J

JOIN—Combines rows from two tables

JSON_AGG()—Aggregates values as a JSON array

JSON_BUILD_ARRAY()—Builds a JSON array

JSON_BUILD_OBJECT()—Builds a JSON object

JSON_EXTRACT_PATH()—Extracts a JSON value at the specified path

JSONB_ARRAY_ELEMENTS()—Expands a JSON array to rows

JSONB_EACH()—Expands the top-level JSON object to rows

K

KEY—Part of constraint definitions

L

LAG()—Window function that provides access to previous rows

LANGUAGE—Specifies the language of a function

LAST_VALUE()—Window function that returns the last value in an ordered set

LATERAL—Subquery that can reference columns from items that appear earlier in the FROM clause

LEAD()—Window function that provides access to subsequent rows

LEAST()—Returns the smallest value from a list

LEFT JOIN—Returns all rows from the left table and matching rows from the right table

LENGTH()—Returns the number of characters in a string

LIKE—Pattern matching using wildcards

LIMIT—Restricts the number of rows returned

LN()—Returns the natural logarithm of a number

LOCALTIME—Returns the current time

LOCALTIMESTAMP—Returns the current timestamp

LOG()—Returns the logarithm of a number

LOWER()—Converts a string to lowercase

M

MAKE_DATE()—Creates a date from year, month, and day values

MAKE_INTERVAL()—Creates an interval value

MAKE_TIME()—Creates a time value

MAKE_TIMESTAMP()—Creates a timestamp value

MAKE_TIMESTAMPTZ()—Creates a timestamp with time zone value

MAX()—Returns the maximum value in a set of values

MD5()—Calculates the MD5 hash of a string

MIN()—Returns the minimum value in a set of values

MOD()—Returns the remainder of a division operation

MODE()—Returns the most frequent value in a set

N

NEXTVAL()—Returns the next value from a sequence

NOT—Negates a condition

NOT IN—Tests if a value doesn't match any value in a list or subquery

NOW()—Returns the current date and time

NTH_VALUE()—Window function that returns the nth value in a window frame

NULLIF()—Returns NULL if two expressions are equal; otherwise returns the first expression

NUMERIC—Exact numeric data type with specified precision and scale

O

OFFSET—Skips rows before starting to return rows

ON—Specifies JOIN conditions

ON CONFLICT—Specifies alternative action for unique constraint violations

OR—Logical operator that returns true when either condition is true

ORDER BY—Sorts the result set

P

PARTITION BY—Divides the result set into partitions for window functions

PERCENTILE_CONT()—Calculates a continuous percentile

PERCENTILE_DISC()—Calculates a discrete percentile

PG_SLEEP()—Pauses execution for a specified number of seconds

PI()—Returns the value of pi

POSITION()—Finds the position of a substring within a string

POWER()—Raises a number to a specified power

PRIMARY KEY—Defines a primary key constraint

Q

QUERY—Used in various contexts to refer to an SQL statement

R

RADIUS—Geographic distance functions

RANDOM()—Returns a random number

RANK()—Assigns a rank to each row

REAL—Approximate numeric data type with single precision

RECURSIVE—Used with CTEs to define recursive queries

REFERENCES—Defines a foreign key constraint

REFRESH MATERIALIZED VIEW—Updates a materialized view

REGR_AVGX()—Returns the average of the independent variable

REGR_AVGY()—Returns the average of the dependent variable

REGR_COUNT()—Returns the number of input rows

REGR_SLOPE()—Returns the slope of the linear regression line

RETURN—Returns from a function

RETURNING—Returns values from inserted, updated, or deleted rows

RIGHT JOIN—Returns all rows from the right table and matching rows from the left table

ROLLBACK—Rolls back the current transaction

ROUND()—Rounds a number to a specified precision

ROW_NUMBER()—Assigns a unique number to each row

S

SAVEPOINT—Creates a save point within a transaction

SCALE()—Returns the scale of a numeric value

SCHEMA—Namespace that contains named database objects

SELECT—Retrieves data from a database

SEQUENCE—Object that generates sequential numbers

SERIAL—Auto-incrementing integer data type

SESSION_USER—Returns the session user name

SET—Changes runtime parameters

SET CONSTRAINTS—Sets constraint checking modes

SETVAL()—Sets a sequence's current value

SHA1(), SHA224(), SHA256(), SHA384(), SHA512()—Cryptographic hash functions

SHOW—Displays the value of a runtime parameter

SIGN()—Returns the sign of a number

SIN()—Returns the sine of a number

SINH()—Returns the hyperbolic sine of a number

SMALLINT—Integer data type with small range

SOME—Returns true if some subquery values meet the condition

SQRT()—Returns the square root of a number

STDDEV()—Returns the standard deviation of a numeric column

STDDEV_POP()—Population standard deviation

STDDEV_SAMP()—Sample standard deviation

STRING_AGG()—Concatenates values with a separator

SUBSTRING()—Extracts a substring

SUM()—Calculates the sum of a set of values

T

TABLE—Refers to a table in the database

TAN()—Returns the tangent of a number

TANH()—Returns the hyperbolic tangent of a number

TEXT—Variable-length character data type

THEN—Executes when a condition is true in a CASE statement

TIME—Time data type

TIMESTAMP—Date and time data type

TIMESTAMPTZ—Timestamp with time zone data type

TO_CHAR()—Converts a value to a string with formatting

TO_DATE()—Converts a string to a date

TO_JSON()—Converts a value to JSON

TO_NUMBER()—Converts a string to a number

TO_TIMESTAMP()—Converts a string to a timestamp

393

TRANSACTION—A unit of work performed within a database system

TRANSLATE()—Replaces specified characters in a string

TRIGGER—Database object that automatically executes in response to events

TRIM()—Removes specified characters from both ends of a string

TRUE—Logical true value

TRUNC()—Truncates a number or date

TRUNCATE—Removes all rows from a table quickly

U

UNION—Combines result sets of two or more SELECT statements (removing duplicates)

UNION ALL—Combines result sets of two or more SELECT statements (keeping duplicates)

UNIQUE—Enforces uniqueness of values in a column

UPDATE—Modifies existing rows in a table

UPPER()—Converts a string to uppercase

USER—Returns the current username

USING—Specifies JOIN columns in a JOIN clause

UUID—Universally Unique Identifier data type

UUID_GENERATE_V4()—Generates a random UUID

V

VACUUM—Garbage collection and optional analysis of a database

VALUES—Constructs table rows from explicitly specified values

VARCHAR—Variable-length character data type

VARIANCE()—Returns the variance of a numeric column

VAR_POP()—Population variance

VAR_SAMP()—Sample variance

VIEW—Virtual table based on a SELECT query

W

WHEN—Specifies a condition in a CASE statement

WHERE—Filters rows based on a condition

WIDTH_BUCKET()—Returns the bucket number into which the operand would be assigned

WINDOW—Named window specification for window functions

WITH—Introduces common table expressions (CTEs)

X

XML—XML data type

XMLAGG()—Aggregates XML values

XMLELEMENT()—Creates an XML element

XMLFOREST()—Creates XML elements from columns

XMLPARSE()—Parses a string as XML

XMLROOT()—Changes properties of an XML value

XMLSERIALIZE()—Converts XML to a string

XMLTABLE()—Produces a table from XML

Y-Z

YEAR—Extracts the year from a date

ZERO()—Various functions that set or test for zero values

Index

A

C

GPSR Compliance
The European Union's (EU) General Product Safety Regulation (GPSR) is a set
of rules that requires consumer products to be safe and our obligations to
ensure this.

If you have any concerns about our products, you can contact us on

ProductSafety@springernature.com

In case Publisher is established outside the EU, the EU authorized
representative is:

Springer Nature Customer Service Center GmbH
Europaplatz 3
69115 Heidelberg, Germany